Lecturing in Geography

Geography Discipline Network (GDN)

Higher Education Funding Council for England
Fund for the Development of Teaching and Learning

Dissemination of Good Teaching, Learning and Assessment Practices in Geography

Aims and Outputs

The project's aim has been to identify and disseminate good practice in the teaching, learning and assessment of geography at undergraduate and taught postgraduate levels in higher education institutions.

Ten guides have been produced covering a range of methods of delivering and assessing teaching and learning:

- Teaching and Learning Issues and Managing Educational Change in Geography
- Lecturing in Geography
- Small-group Teaching in Geography
- Practicals and Laboratory Work in Geography
- Fieldwork and Dissertations in Geography
- Resource-based Learning in Geography
- Teaching and Learning Geography with Information & Communication Technologies
- Transferable Skills and Work-based Learning in Geography
- Assessment in Geography
- Curriculum Design in Geography

A resource database of effective teaching, learning and assessment practice is available on the World Wide Web, http://www.chelt.ac.uk/gdn, which contains national and international contributions. Further examples of effective practice are invited; details regarding the format of contributions are available on the Web pages. Examples should be sent to the Project Director.

Project Team

Lead site: *Cheltenham & Gloucester College of Higher Education*
Professor Mick Healey; Dr Phil Gravestock; Dr Jacky Birnie; Dr Kris Mason O'Connor

Consortium: *Lancaster University*
Dr Gordon Clark; Terry Wareham
Middlesex University
Ifan Shepherd; Professor Peter Newby
Nene — University College Northampton
Dr Ian Livingstone; Professor Hugh Matthews; Andrew Castley
Oxford Brookes University
Dr Judy Chance; Professor Alan Jenkins
Roehampton Institute London
Professor Vince Gardiner; Vaneeta D'Andrea; Shân Wareing
University College London
Dr Clive Agnew; Professor Lewis Elton
University of Manchester
Professor Michael Bradford; Catherine O'Connell
University of Plymouth
Dr Brian Chalkley; June Harwood

Advisors: Professor Graham Gibbs (*Open University, Milton Keynes*)
Professor Susan Hanson (*Clark University, USA*)
Dr Iain Hay (*Flinders University, Australia*)
Geoff Robinson (*CTI Centre for Geography, Geology and Meteorology, Leicester*)
Professor David Unwin (*Birkbeck College, London*)
Dr John Wakeford (*Lancaster University*)

Further Information

Professor Mick Healey, Project Director   Tel: +44 (0)1242 543364   Email: mhealey@chelt.ac.uk
Dr Phil Gravestock, Project Officer   Tel: +44 (0)1242 543368   Email: pgstock@chelt.ac.uk
Cheltenham & Gloucester College of Higher Education
Francis Close Hall, Swindon Road, Cheltenham, GL50 4AZ, UK   [Fax: +44 (0)1242 532997]

http://www.chelt.ac.uk/gdn

Lecturing in Geography

Clive Agnew and Lewis Elton

University College London

Series edited by Phil Gravestock and Mick Healey
Cheltenham & Gloucester College of Higher Education

Published by:

Geography Discipline Network (GDN)
Cheltenham & Gloucester College of Higher Education
Francis Close Hall
Swindon Road
Cheltenham
Gloucestershire, UK
GL50 4AZ

Lecturing in Geography

ISBN: 1 86174 025 5 ✓
ISSN: 1 86174 023 9

Typeset by Phil Gravestock

Cover design by Kathryn Sharp

Printed by:

Frontier Print and Design Ltd.
Pickwick House
Chosen View Road
Cheltenham
Gloucestershire, UK

Contents

Editors' preface

This Guide is one of a series of ten produced by the Geography Discipline Network (GDN) as part of a Higher Education Funding Council for England (HEFCE) and Department of Education for Northern Ireland (DENI) Fund for the Development of Teaching and Learning (FDTL) project. The aim of the project is to disseminate good teaching, learning and assessment practices in geography at undergraduate and taught postgraduate levels in higher education institutions.

The Guides have been written primarily for lecturers and instructors of geography and related disciplines in higher education and for educational developers who work with staff and faculty in these disciplines. For a list of the other titles in this series see the information at the beginning of this Guide. Most of the issues discussed are also relevant for teachers in further education and sixth-form colleges in the UK and upper level high school teachers in other countries. A workshop has been designed to go with each of the Guides, except for the first one which provides an overview of the main teaching and learning issues facing geographers and ways of managing educational change. For details of the workshops please contact one of us. The Guides have been designed to be used independently of the workshops.

The GDN Team for this project consists of a group of geography specialists and educational developers from nine old and new universities and colleges (see list at front of Guide). Each Guide has been written by one of the institutional teams, usually consisting of a geographer and an educational developer. The teams planned the outline content of the Guides and these were discussed in two workshops. It was agreed that each Guide would contain an overview of good practice for the particular application, case studies including contact names and addresses, and a guide to references and resources. Moreover it was agreed that they would be written in a user-friendly style and structured so that busy lecturers could dip into them to find information and examples relevant to their needs. Within these guidelines the authors were deliberately given the freedom to develop their Guides in their own way. Each of the Guides was refereed by at least four people, including members of the Advisory Panel.

The enthusiasm of some of the authors meant that some Guides developed a life of their own and the final versions were longer than was first planned. Our view is that the material is of a high quality and that the Guides are improved by the additional content. So we saw no point in asking the authors to make major cuts for the sake of uniformity. Equally it is important that the authors of the other Guides are not criticised for keeping within the original recommended length!

Although the project's focus is primarily about disseminating good practice within the UK a deliberate attempt has been made to include examples from other countries, particularly North America and Australasia, and to write the Guides in a way which is relevant to geography staff and faculty in other countries. Some terms in common use in the UK may not be immediately apparent in other countries. For example, in North America for 'lecturer' read 'instructor' or 'professor'; for 'staff' or 'tutor' read 'faculty'; for 'postgraduate' read 'graduate'; and for 'Head of Department' read 'Department Chair'. A 'dissertation' in the

UK refers to a final year undergraduate piece of independent research work, often thought of as the most significant piece of work the students undertake; we use 'thesis' for the Masters/ PhD level piece of work rather than 'dissertation' which is used in North America.

In addition to the Guides and workshops a database of good practice has been established on the World Wide Web (http://www.chelt.ac.uk/gdn). This is a developing international resource to which you are invited to contribute your own examples of interesting teaching, learning and assessment practices which are potentially transferable to other institutions. The resource database has been selected for *The Scout Report for Social Sciences*, which is funded by the National Science Foundation in the United States, and aims to identify only the best Internet resources in the world. The project's Web pages also provide an index and abstracts for the *Journal of Geography in Higher Education*. The full text of several geography educational papers and books are also included.

Running a consortium project involves a large number of people. We would particularly like to thank our many colleagues who provided details of their teaching, learning and assessment practices, many of which appear in the Guides or on the GDN database. We would also like to thank, the Project Advisers, the FDTL Co-ordinators and HEFCE FDTL staff, the leaders of the other FDTL projects, and the staff at Cheltenham and Gloucester College of Higher Education for all their help and advice. We gratefully acknowledge the support of the Conference of Heads of Geography Departments in Higher Education Institutions, the Royal Geographical Society (with the Institute of British Geographers), the Higher Education Study Group and the *Journal of Geography in Higher Education*. Finally we would like to thank the other members of the Project Team, without them this project would not have been possible. Working with them on this project has been one of the highlights of our professional careers.

Phil Gravestock and Mick Healey

Cheltenham

July 1998

All World Wide Web links quoted in this Guide were checked in July 1998. An up-to-date set of hyperlinks is available on the Geography Discipline Network Web pages at:

http://www.chelt.ac.uk/gdn

About the authors

Clive Agnew

I am an applied climatologist and senior lecturer in physical geography at University College London. My first degree was in Geography (at Newcastle), I then undertook a PhD in Development Studies (1980) at the University of East Anglia and worked on systems analysis at the Open University. My research has focused upon Africa and the Middle East on topics of water resources, drought and environmental degradation. Recent studies include surface runoff collection in the Badia region of Jordan, rural water supplies in Uganda and the problem of assessing drought and desertification impacts in drylands. I am currently working on drought and low flows in Southern England and developing an operational model for wetland evaporation assessment. I have written teaching guides for the Open University and the University of London and have just (1998) been given the Distinguished Teacher Award at UCL.

Lewis Elton

Since 1994 I have been Professor of Higher Education at University College London and I am also Professor Emeritus of Higher Education, University of Surrey. My academic career has been in two parts, first as a theoretical physicist (I was Professor of Physics and Head of Department at the University of Surrey, 1962-1971 and am a Fellow of the American Institute of Physics); and then in higher education (I am a Member of the Staff and Educational Development Association, and I was a Higher Education Adviser to the Employment Department, 1989-1994). I also hold a Doctor of Letters (honoris causa) of the University of Kent at Canterbury. In the past ten years my research has been mainly in the management of higher education, before that it was in curriculum development and in teaching and learning in higher education. Other recent work has included chairing a Task Force on 'Staff Development in Relation to Research'; evaluating the Quality Assessment Programme of the Higher Education Funding Council for Wales and developing a Postgraduate course on 'Research and Development in Higher Education' for experienced academic teachers.

1   Introduction to this Guide

1.1   Why bother with lectures?

What is a lecture? In the context of this project there are separate Guides on small class teaching, for example Clark & Wareham (1998), Birnie & Mason O'Connor (1998) and Livingstone *et al.* (1998). Lectures are taken here to be concerned with the teaching of large classes in a conventional lecture theatre or large room. This does not mean however that the suggestions given below cannot be employed with smaller groups or in other venues such as fieldwork or in the laboratory. It makes little difference whether the lecture is to 5 or 500, the principles remain the same (Cryer & Elton, 1992). Cox (1994) distinguishes between lectures and role playing, independent learning, peer teaching, private study and computer assisted learning. We will not be using so rigid a classification, as student participation can enhance a lecture while what happens after a lecture is just as important as its delivery and content. There are many reasons for deciding to lecture:

- They are a very cost effective means of teaching large classes.
- They are useful when bodies of factual information need to be presented to students.
- They enable teachers and students to organise their time effectively.
- They enhance control over class content, facilitating curriculum and study-programme planning.
- They are an efficient use of lecturer time as once prepared they are quickly updated and reusable.
- They can lead to personal satisfaction.
- They introduce students to the language of geography (the spoken and written word).
- They can impart enthusiasm for the subject.
- They promote basic cognitive skills such as memorizing facts.
- They allow key concepts and principles to be promoted.

(Note that these reasons can also apply to other teaching methods)

How many of these reasons applied to the last lecture you gave? Are there other reasons why you lecture that are not included in the above list?

Brown & Atkins (1988) trace the use of lectures to Ancient Greece, five centuries before Christ, and note that the term is derived from medieval Latin *lectura* meaning **reading aloud**. However, to make reading aloud sound spontaneous is an actor's skill which few lecturers have mastered; unless that is the case, most lecturers do better by rehearsing their lectures well and then presenting them, using notes to prompt the memory. A very few can deliver outstanding lectures without such prompts and as if spontaneous (which they are not); the noted historian A.J.P. Taylor was one of these.

In essence, a lecture is characterised by the lecturer being in control. The lecturer determines the content, the pace of delivery and the organisation of material and ideas. In more innovative styles of teaching, control is shared between staff and students through greater audience participation. It follows that a distinction can be made between **formal** and **interactive lectures**. A formal lecture is characterised as 'chalk and talk', an uninterrupted address, often supported by visual aids, but with few questions and often abundant student note taking. Gold *et al.* (1991, p.8) suggest that:

> *'the typical lecture is a one way transmission of information which takes place at regular (probably weekly) intervals and lasts 50 to 55 minutes.'*

It is widely agreed that 'typical lectures' tax the attention span of students and means should be found to shorten and break up the presentation, even if not going as far as making the lecture genuinely interactive.

An interactive lecture, by contrast, includes periods when the students participate in the presentation and/or discussion of ideas and information, for example, through buzz groups, brainstorming or demonstrations. Whether formal or interactive, lectures should always be followed up with supplemental activities such as student reading or tutorial discussions and these are at least as important to student learning as the 50 minute delivery. However, such supplemental activities usually depend for their success on good note taking in the lectures, of which more in Section 3.

Figure 1 (from Brown & Atkins, 1988) shows the traditional perspective of teaching in Higher Education (HE) as a trade-off between lecturer versus student participation and control. The ideal is a greater shared responsibility, particularly when using lectures, and 'interactive lectures' should be placed somewhere between 'lecture' and 'small group teaching'.

Figure 1: *Lecturer controls versus student participation for a range of teaching methods (based on Brown & Atkins, 1988).*

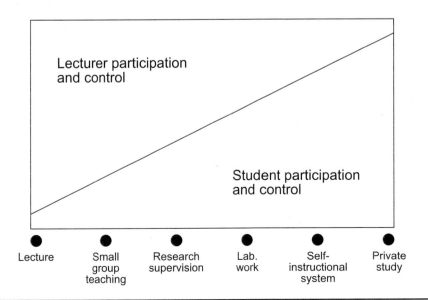

1.2   Why do you need to improve your lectures?

We assume you are reading this Guide because you want to improve your teaching using lectures. This is not surprising. It is highly fashionable to denigrate the lecture as a learning process and we have all been subjected to boring, rambling downright bad lectures. The old joke

> *'A lecture is the means by which the notes of one individual (the lecturer) are transformed into the notes of another individual (the student) without passing through the brain of either'*

is still recounted to new members of staff. Student experience would appear to endorse this view:

> *'It seemed so stupid, just copying and not realising what you are doing.'*

> *'It was like sitting there looking at a television screen: you know, just looking.'*

> *'Soon the chalk is pouring across the blackboard, and he'll talk to the blackboard – he won't talk to – won't even necessarily look at – the students. I feel very anti him, anti what he is doing.'*

Contrast these experiences with the following comments:

> *'You feel mentally stimulated – you feel you can go away and read up on it and be interested in it.'*

> *'There'd be a hush the moment he walked in. Everybody was interested. He got on with the conversation – er lecture – yes conversation is the right word – it was as if he was talking to you individually.'*

> *(Bliss & Ogborn, 1977)*

If we are honest we will also admit to having given less than satisfactory lectures. With current pedagogic thinking concerned about 'deep learning', 'student participation' and 'life-long learning' it might appear that lecturing is a dying form of teaching. Nothing could be further from the truth. An inspection of rising student numbers in geography, worsening staff:student ratios and declining units of resource for teaching in HE leads one inexorably to the conclusion that effective ways of teaching large numbers are required.

Table 1:   *Class sizes in geography departments (from Jenkins & Smith, 1993)*

	1986	1991
Year 1 Lecture course	64.6	95.0
Year 2 Lecture course	22.0	29.3

Staff:student ratios have worsened over the same period with a 44% increase from 1:12.1 to 1:17.4

But does this mean that the lecture is merely a cost-effective means of delivery? This Guide hopes to show you that the answer is an emphatic **NO**. As Cox (1994, p.59) states:

> *'Some authors claim that formal lecturing is only appropriate for certain objectives (conveying knowledge, facilitating comprehension), but not for others (applications, analysis and synthesis, evaluation and criticism). I believe that a properly designed and delivered lecture can achieve all of these objectives. Put bluntly, a lecture can achieve anything that can be done by talking to someone.'*

In addition, it can satisfy affective and attitudinal objectives, for are these not the main objectives to be satisfied by sermons and is a sermon not a form of lecture? There is nothing wrong with occasionally preaching as well as teaching in a lecture, **provided it is done well**.

This Guide to lecturing in geography will provide you with some ideas for changing your lecturing style and methods of delivery, but the 'performance' is only part of the challenge. We also hope to encourage you to think about the aims of your lecture in the context of student learning and hence to become a more effective teacher. After all your time is valuable and it is in everybody's interest, especially the student's, that lectures achieve as much as possible.

The emphasis here is on practical advice. We realise that the pressures on staff in HE are such that strategies that are expensive in terms of staff time and facilities are liable to be ignored. However, it is rare to get something for nothing and improving your teaching will have an initial cost of your time at the very least. Here are possible reasons for improving your lectures:

- Difficulties of dealing with increasingly large classes of students.

- Critical comments from student evaluations of your course.

- Pressures from peer review of teaching.

- Teaching quality assessment is imminent.

- Teaching is taken into account for promotion.

Take time to add one or two reasons of your own to this list

1.

2.

3.

These are all 'real' reasons but they are also all criticisms or threats and may well push you into adopting methods that are perceived as being a 'quick-fix' based on 'current good practice'. The danger is that the student learning experience will not necessarily improve. It is hoped then that some of the following reasons for improving your lectures will also apply:

- Personal satisfaction of delivering interesting and motivating lectures.

- Desire to encourage students to think for themselves.

- Dissatisfaction with the passive role of students in lectures.

1.3   What are the aims of this Guide?

Bligh (1971) in his useful book on 'What's the use of lectures?' noted that anyone offering advice to a colleague about how to improve their lecture is likely to be ridiculed the next time he or she addresses a group of students as there are very few virtuoso performers, but then this is equally true of violinists or footballers who, whether virtuosos or not, all aim to improve through training. The authors of this Guide would never claim to embody perfection when it comes to lectures. The aim here is to bring evidence of good practice to your attention with practical suggestions and advice on possible implementation.

The aims of this Guide are then:

- To promote good teaching practice, including, particularly, the incorporation of active learning in lectures.

- To encourage innovations and new lecturing methods.

- To summarise and present geography teachers' attempts to improve teaching by lectures.

The criticisms levied against the traditional lecture, even if delivered well, relate to it as a learning device. We will therefore be encouraging you to think about ways in which students can become more directly involved in the lecture and therefore contribute to and gain more from this learning process.

2   What is wrong with lectures?

There are three types of problems concerning lectures in HE:

- Lectures are far too common and there is a need for greater diversity with different types of challenge for teacher and student.

- The recent increase in class sizes, institutional changes and widening access to students present a number of challenges.

- Lectures offer limited opportunities for student's learning and may achieve very little when the audience merely acts as a passive observer.

2.1   Are there too many lectures?

The lecturing mode continues to dominate teaching in HE. An analysis in 1988 of 39 replies from UK geography departments on the teaching of geomorphology revealed that for every one hour spent in the laboratory 2.2 hours were spent in the lecture theatre (Petch & Reid, 1988). For environmental science departments the ratio rose and for every 1 hour spent in the laboratory 4.2 hours were spent in lectures. This is in a subject where laboratory skills and fieldwork training is seen as being increasingly important and yet Bishop *et al.* (1995, p.104) still assert that,

> *'Many geomorphology courses are lectures without laboratories,'*

More recently Phillips & Healey (1996) examined the teaching of the 'History and Philosophy of Geography in Great Britain' and found that lectures were still prevalent. They noted a wide range of teaching methods and much change in recent years with a trend towards small group teaching and individual learning; yet 35% of teachers suggested their students spent over three quarters of their time in lectures and 65% replied that students were lectured to in at least half the classes.

> How much of your teaching is performed using lectures as opposed to other types of presentation, do lectures account for more than half of your taught classes?

An over-dependence upon lectures for teaching in geography departments was also noted in the recent Teaching Quality Assessment. Healey (1997, p.99) reported that HEFCE (1995) listed amongst the areas for improvement:

> *'An over-dependence on lectures and the need for diversity and innovation,'*

and recommended that departments:

> *'Introduce greater challenge and interaction into that small proportion of lecture classes which, whilst technically sound, were sometimes staid and unchallenging.'*

The continued domination of lectures as a method of teaching is explained in part by the decline in units of resource for teaching geography students (Gibbs & Jenkins 1992; Jenkins 1997) and a commensurate increase in class size. This has resulted in cost effective methods of teaching large numbers of students being favoured at the expense of more innovative or resource-demanding approaches. The recent changes taking place in HE and their implications are discussed in the GDN Guide by Gardiner & D'Andrea (1998), but it is inappropriate for any guide on 'good practice' to ignore the major constraints confronting all HE teachers. Crewe (1996) summarises these as being:

- A rise in student numbers.

- Inadequate resources.

- Mounting pressures on staff time.

- Increasing competition for research grants.

- The casualisation of employment.

To this list we can add the introduction of a range of measures of quality control, modularisation, semesterisation with two examination periods, and the opening up of access to HE to a wider range of students. But the greatest impact on teaching is the diminution of resources for teaching coupled with a rise in student numbers which leads to larger class sizes. Gibbs *et al.* (1996) analysed teaching at Oxford Brookes University as an example of an institution where these changes had been experienced between 1984 and 1994. Over this period the maximum class size increased from 65 to 113 students, while nationally student numbers in geography rose by 43% between 1986 and 1991 (see also Table 1, p.4). These changes present you with a number of challenges illustrated by the following quotes from geography teaching staff (Jenkins & Smith, 1993):

> *'There is the dilemma between the less effective method of coping by lecturing to large numbers or by reducing contact time and having small groups.'*

> *'We are adjusting to large groups with more formal lectures and fewer small groups.'*

> *'Staff have worked harder and longer. It is apparent this cannot persist much longer...The department is beginning to adjust to a change of practice that is necessary with the continued rise in numbers. It is clear that the tried and trusted methods no longer suffice.'*

2.2 Do you know your students?

The increase in student numbers includes a widening of access for students so that there are now more mature students and more students with non-traditional entry qualifications. These changes in access pose both problems and opportunities. In a large class it can no longer be assumed that students have similar backgrounds. Their knowledge base is likely to be more variable in the future. This diversity is further accentuated where modularisation of degree courses has been introduced and students in any one class may be majoring in a range of subjects. These changes present opportunities as students generally have greater experience and a wider range of transferable skills.

How have the students you teach changed in recent years? Do you know the background of the students you teach? What information is easily available to you about the students in your class?

Do you know your students?

Lewis Elton of the Higher Education Research Development Unit (HERDU) at University College London (UCL) in his inaugural lecture at UCL challenged his audience of University teachers to decide whether they were teaching a dromedary or a bactrian camel. He wasn't talking to a room of zoologists but was referring to whether a student audience has one hump or two? If you are still wondering what camels and humps has to do with geography students then consider your own specialist subject area. Take any first year class, what is their understanding of your subject? In terms of their understanding are most students grouped around a mean with some knowing very little and a few very well informed? Are there several groups, each with different backgrounds and levels of competence? If it is the latter what are the implications for both the content and style of your lecture? How can you decide upon the appropriate content and starting standard?

One solution is to move away from an emphasis upon content and towards an emphasis upon student learning. But even designing an appropriate teaching strategy will require you to make assumptions about your students' ability and knowledge when it is quite likely that you do not know enough about either, especially for a large first year class. In England the Universities and Colleges Admissions Service (UCAS) entry form only provides limited information on a student:

Their academic record including examination board

Home address

Age

Gender

but there may be many things you would like to know. The entry questionnaire on the opposite page is used at UCL to elicit more useful information.

Incoming students questionnaire
Geography, UCL

Questions are asked on the following topics:

General information:

        Type of school/college
        Gap year or not
        Single sex/mixed
        Why chose UCL

Geography teaching:

        Which geography A-level (or other form of pre-university study)
        Main topics covered in geography classes
        Size of group in which taught
        Were essays written (geography/other subjects)
        Roughly how long (number of pages) were the essays
        How many essays
        Was a project (independent study) completed
        Title of the project work
        Project preparation
        Geography fieldwork
        Residential fieldwork
        Key text books used
        Use of photocopied materials
        Reading of geography journals
        Expectations of UCL geography course

Transferable skills:

        Group work
        Oral presentations
        Library research
        Use of WWW
        Use of email
        Use of word processing
        Use of statistical analysis packages
        Other IT
        What skills are hoped to be gained while at UCL

(for further information on the questionnaire please contact Dr Ann Varley, avarley@geography.ucl.ac.uk)

2.3   Can students learn in lectures?

The analysis by the Higher Education Learning Project (Bliss & Ogborn, 1977) found that nearly half of all 'stories' from students concerned lectures rather than other types of teaching. Two thirds of these were critical, but one third was positive. Students who praised lectures often mentioned increased self-confidence linked to greater understanding, or stimulation of interest producing higher motivation to study. Those who were critical either felt a sense of personal failure and inadequacy, or they felt antagonistic towards the lecturer. This was explained in terms of lack of preparation for the lecture, inability to communicate, being brusque or not appearing to understand the material presented. For many no blame was attached but the student simply withdrew from the learning process, switching off or not bothering to turn up.

What do you hope to achieve in your lectures? Make a list in order of importance of what you achieved in the last lecture you gave. How many of these objectives were concerned with content and how many required the students to think about the material presented during the lecture?

1.

2.

3.

4.

5.

Bligh (1971) noted three possible achievements of lectures:

1.   The acquisition of information.

2.   The promotion of thought.

3.   Changes in attitude.

The last of these are only achievable when presented by particularly gifted public speakers. Gold *et al.* (1991) reported lectures are often traditional and formal, overly factual and descriptive with little analysis, rarely exciting for students, and promoted considerable note taking with an emphasis upon examination preparation. A passive process for all involved. Bligh went on to argue (p.25),

> '*Although it is sometimes believed that the lecture method can fulfil three kinds of function, the available evidence suggests that it can only effectively achieve one — the student's acquisition of information…. It is therefore suggested that new teachers should use the lecture method primarily for this purpose.*'

Do you agree with this assertion, does it apply to your lectures?

Figure 2:   *The different contents of lectures*

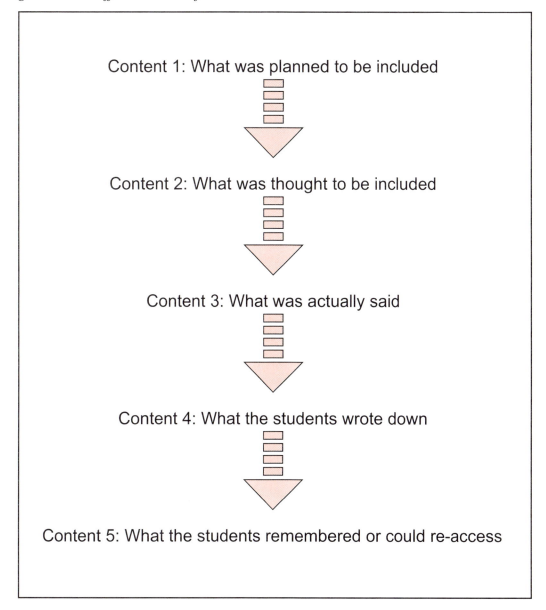

Content 1: What was planned to be included

Content 2: What was thought to be included

Content 3: What was actually said

Content 4: What the students wrote down

Content 5: What the students remembered or could re-access

Are lectures able to convey information successfully? Figure 2 suggests there are many opportunities for information to be lost or confused from lecture presentation to student note taking and afterwards. The view that the lecture is useful for the dissemination of information is no longer widely accepted and Gould (1994, p.277) reflecting upon geographical thinking remarked that

> *'teaching well is one of the most difficult things in the world, and we all recognise the difference between teaching that is anecdotal entertainment, and that which opens up new ways of seeing...(he continues)...my initial feelings of fraudulence (as a teacher — added by author) were heightened by a long standing and unresolved worry about teaching as purveying facts to be memorised, or teaching as opening up ways of thinking.'*

Entwistle (1984) listed a number of learning outcomes expected by lecturers but found one unifying theme, the aim to promote

> *'Critical thinking'.*

More recently, in Australia, Isaacs (1994) found the same response. Lectures which concentrate on increasing students' knowledge fail in the promotion of critical thinking on two grounds:

- there is no time left for it;

- the attitudes engendered by the usually authoritarian approach to increasing students' knowledge is inimical to its development.

The first must be dealt with by ruthless pruning. The second by a more reflective style of lecturing with the lecturer acting as a guide and facilitator. That is the switch from surface to deep learning.

2.3.1   Deep versus surface learning

There is much discussion over the merits of 'surface' versus 'deep' learning (Cryer & Elton 1992). In surface learning new knowledge is imparted by the lecturer, but this does not lead to greater understanding and encourages regurgitation by students. Deep learning involves the integration of new knowledge with what is already known, leading to greater understanding. This necessitates that students reflect on their learning, for which there is normally no time in formal lectures, and it cannot be stressed too strongly that even if you present your material in a way that is suitable for deep learning, students may receive it — if only for lack of reflecting time — as surface learning. It follows that the formal lecture usually promotes surface learning while interactive lectures can promote deep learning. Students need time to reflect and think; this may commence in the time allotted to the lecture but certainly should be expected to continue thereafter. Unfortunately, once students have used surface learning in the lecture and in their note taking, they find it very difficult to switch to deep learning afterwards. The result is that they come to tutorials in order to learn what they ought to have learned in lectures and the frequently observed repetition of parts of a lecture several times in tutorial groups is probably the most inefficient use of a teacher's time.

When switching from lecturing to self-instructional materials, Lewis Elton (author) immediately found that students no longer used tutorials to have the material repeated for them. This indicates that **good** notes are important for deep learning. It also enabled Lewis to say 'Did you talk it over with your fellow students before you came to me?'.

See Sections 3 and 4 in this Guide for examples that promote student reflection both during the lecture and afterwards.

2.3.2   Active learning

This involves students in some activity during the lecture in order to promote deep learning through participation, discussion and reflection. Many lecturers readily recognise the limitations of the formal lecture in facilitating active learning. Gold *et al.* (1991) describe the process from passive to active learning as moving from receiving knowledge, to exploring existing knowledge, to students creating their own knowledge.

Figure 3:   *Student concentration and learning loss in lectures (after Cox, 1994)*

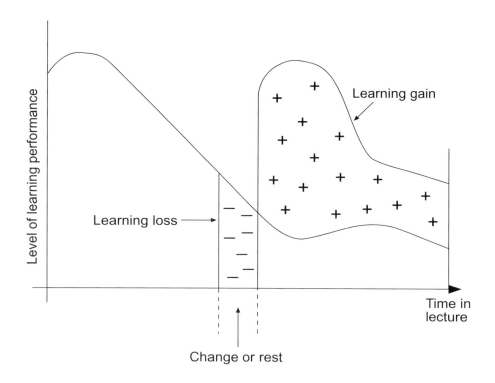

Active learning is more than merely taking a break to enhance attention, as it involves thinking about what has been presented. Students also have a responsibility to participate in either form of delivery but Matthews & Livingstone (1996) argue that the learning objectives and students' role in the process have to be clearly explained.

It should be borne in mind that if time is devoted to student participation then there is inevitably less available for imparting facts and ideas (Cryer & Elton, 1992). Adoption of active learning consequently results in a reduction of lecture content. The emphasis changes from imparting material to one of guidance. But deep learning is not achieved simply by involving students in a number of lecture-based activities such as buzz groups. Getting the students to interact with the lecturer, for example, using questions and answers, especially with large classes, is merely the first obstacle to be overcome. The move from formal to interactive lecturing should have one further consequence — it should demote lecturing from its primacy in university teaching. Lectures — and there should certainly be fewer of them — now become just one of many parts of an integrated learning system, which should include resource-based learning, independent learning, learning through peer teaching, practical and project work.

3 Lecture methods

Gibbs (1981, p.4) lists several explanations for the continued use of the traditional, formal lecture including:

Lectures are the best way to get facts across.

Lectures make sure that students have a proper set of notes.

The criticisms one can make of lecturing only apply to bad lecturing.

It has been argued here that at best a formal lecture provides enthusiasm for the subject matter, invokes some critical thinking and imparts an amount of knowledge, but these are rarely achieved in practice, as is a 'proper set of notes'. We are assuming that you want to improve your lectures; whether this will embrace active learning or whether you wish to keep to traditional modes of delivery will be a personal decision based on what you are hoping to achieve with your students.

This section presents practical advice on a wide range of lecturing issues that can be augmented by some of the numerous texts on this topic (Brown & Atkins, 1988; Cox, 1994, Gibbs & Jenkins, 1992; Gold *et al.*, 1991; Hay, 1994 and Race & Brown, 1994). It is worth noting at the outset the potential costs involved and problems of transferability. Just because an approach works in one context does not mean it is wholly transferable to others. Modularisation, semester teaching, library resources, computing facilities and the availability of teaching assistants may affect the success or failure of an approach. Perhaps most important of all you need your colleagues and institution to agree that the time you are about to invest in improving your lectures is well spent.

Here are number of reasons why you may find it difficult to change your teaching. Examine the list and identify which of these apply to you, select those reasons you can do something about and decide a plan of action:

- Ignorant of the alternatives.
- Do not know how effective lectures are.
- Overworked.
- Changes take time.
- The alternatives to lectures appear to involve more work.
- There is a lack of resources.
- The institution supports lectures.
- External validation.

3.1   Delivery and presentation

There are many forms of delivery from the eclectic to the amorphous speaker (see Brown & Bakhtar, 1988). Whatever style is adopted there are several tasks that must be completed. You must grab the attention of your audience at the outset. The introduction should be particularly interesting and lively as first impressions are important. You must be lucid and audible throughout the room. You must establish a rapport with the audience through the projection of your personality. This is not to advocate we all adopt an extrovert 'over-the-top' style of presentation; a reflective, considered and gently paced lecture can gain more respect and attention than one from someone who bounds around the room gesticulating wildly. Non-verbal behaviour such as smiling or making eye contact is also important, although too much movement can be a distraction. Most if not all of these points should be included in staff training where play-back of video tapes of one's own lectures is an effective learning process.

Which of the following criticisms of presentations by lecturers (after Brown & Atkins, 1988, p.13) apply to your last presentation?

Students' views	Lecturer's self criticism
Inaudible	Saying too much too quickly
Incoherent	Assuming too much knowledge
Inappropriate level of content	Forgetting to summarise
Not emphasising key points	Not indicating asides (digressions)
Poor visual presentation	Poor timing

Lack of clarity appears to be a main failing and the same authors propose the following should be adopted in order of priority (p.22):

Speaking: Speak clearly, do not speak too fast and use pauses.

Planning: Plan, prepare and structure the material which should simplify the topic (not make it simplistic).

Clarifying: Make it understandable and clarify key points.

Observing: Observe student reactions.

Omitting: Do not try to include everything.

Checking: Check you understand your own material.

Saroyan & Snell (1997) analysed three types of lecturing style:

Lecture 1: Content driven, covered a large amount of information and followed a systematic structure with 7 different topics.

(cont.)

Lecture 2: Commenced with three case studies and then covered fewer topics, only three in this instance. Students had an opportunity to discuss the information and the learning outcomes were explicitly stated.

Lecture 3: Four topics were covered but less time was devoted to lecturing with equal time given to answering students questions.

Student evaluation favoured lecture 3 style with lecture 1 receiving lowest scores although lecture 2 was rated highest for overall teaching effectiveness. This does not mean one lecturing style yielded better learning but suggests that students preferred a lecturing style that provided the opportunity to apply knowledge gained during the lecture. The authors note that the effectiveness of a lecture can be assessed by consideration of:

Appropriateness?

Is lecturing appropriate for the desired learning outcome? For example, lecturing has been shown to be an effective strategy when the objective is for students to acquire knowledge of facts.

Organisation?

Does the lecture enable students to make the links between the concepts presented?

Interactiveness?

Are students engaged in activities which are relevant to the intended learning?

3.2   Content and structure

Gold *et al.* (1991, p.11) recount the occasion when an eminent historical geographer (H.C. Darby) could not present his lecture and so arranged for a colleague (Terry Coppock) to use his notes as a replacement with the following results:

> *During the first half hour of the lecture no one took any notes but after about 30 minutes the audience picked up their pens and began writing. He later realised that he had given the previous week's lecture for the first half hour.*

We might use this story to point out the need to check with the audience their understanding of the material presented, that is a lecture should be a two-way communication process. But perhaps more pertinent is the observation that it only took the colleague half the time to deliver the same material that the eminent historian required with all his asides, anecdotes and pauses. Most lecturers include too much material, especially slides and overheads. To some extent this shows a lack of confidence, ensuring that the presentation does not finish early or cramming facts in to convince students and oneself that the lecture has been worthwhile. We provide below some suggestions as to how such 'gaps' can be used to good purpose. Pruning the content so that there is more thought and less facts is sometimes a difficult task. We doubt you speak at more than 100 words a minute (check for yourself) so you would only get through 5,000 words in the whole of a lecture even if you spoke continuously.

How many visual aids do you normally allow for a standard 50 minute lecture? If you cut this number by half and spent twice as much time on those remaining how would this improve the quality of the lecture? How else might you reduce the content of your lectures? What could this time be used for?

We have already made the point that the factual content of your lectures is probably too high. The lecture is best used to convey key issues, identify priorities and provide broad outlines. This is only effective with a coherent structure.

Hay (1994, p.58) lists several frameworks used by geographers:

Chronological	e.g. history of geographical thought from the 19th Century.
Spatial	e.g. description of Japan's trading relations with other countries of the Pacific.
Causal	e.g. implications of financial deregulation on the New Zealand insurance market.
Order of importance	e.g. ranked list of solutions to the problem of male homelessness in Adelaide.

Many of these frameworks, such as the spatial, emphasise a hierarchical ordering of ideas moving from main topic to subset, or being problem-centred and moving from causes to solutions. Bligh (1971) suggests there are many other structures including chains of ideas which form a story line or a chronology, to more complex networks where there are complex interrelationships and no simple organisation. A problem for you is that a serial presentation of points may obscure the structure of your ideas, for example see Figure 4.

Figure 4:   *Sequence of presentation of a hierarchic lecture (based on Bligh, 1971)*

Topic — Estimating evaporation rates

Figure 4a: *Topic organisation*

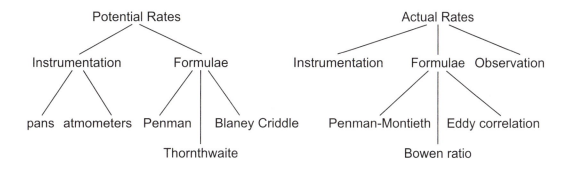

Figure 4b: *Lecture organisation*

1. *Potential Rates*
    A. Instrumentation
        i. pans
        ii. atmometers
    B. Formulae
        i. Penman
        ii. Thornthwaite
        iii. Blaney-Criddle
2. *Actual rates*
    A. Instrumentation
    B. Formulae
        i. Penman-Montieth
        ii. Bowen ratio
        iii. Eddy correlation
    C. Observation

Whatever the structure, Bligh emphasises it should be explained at the outset in order to help students anticipate and organise their own thoughts. But it is equally useful to provide summaries or a progress report, i.e. to refer back to the structure and identify the point you have reached. A terminology has been developed (see Brown & Atkins, 1988; Gold *et al.*, 1991) to explain different ways to do this:

Table 2: *Explaining the lecture structure to students (after Brown & Atkins, 1988; Gold et al., 1991)*

CONCEPT	MEANING	EXAMPLE
Signposts	Statements presenting the structure and content	'Today I want to examine the three main causes of drought: a reduction in water supply, an increase in demand, and mismanagement. In tackling each possible cause I will be comparing the role of nature versus the activities of humans. At the end I will use a case study from Britain and one from Africa to exemplify these different causes.'
Frames	Statements that delimit the beginning and end of sections	'Let us now turn to the topic of the natural causes of drought starting with the reasons for a decline in precipitation.'
Foci	These signal important points	'In 1965 Palmer, and many climatologists since, regarded a lack of rainfall as the prime cause of drought.'

| *Links* | Statements that link to the sections of the lecture together or link to previous talks or students' experiences | 'Note that confusion over the nature of drought is due to many of the same reasons given in last week's lecture on desertification such as vague definitions, multi-faceted causes and institutional facts.' |

Clarity and coherence can also be aided by explaining to students what you expect them to have gained by the end, for example:

> *'at the end of this lecture on accelerated global warming you should understand the role of greenhouse gases, be able to suggest reasons why there is uncertainty over future emissions of these gases and be able to discuss the ways Britain could curb future greenhouse gas emissions.'*

Clarity can be further enhanced by giving students an activity at the end of the lecture to see which of the outcomes they have understood.

> Take time to reflect upon a recent lecture you have given. Did you explain what was expected from the students both during and after the lecture i.e. follow-up? How do you know if they were successful or not without formal assessment?

The conclusion reached so far is that even without changing to interactive lecturing, we could significantly improve the lecture and make it more of a medium for active learning. This would require three measures:

1) A radical pruning of the material to be presented

'Cutting down syllabuses' has proved very difficult, which is due in part at least to lecturers not being prepared to give up any of their material in a teaching system where lecturers act largely independently of each other. A change to a course team approach and an analysis in terms of learning objectives can both help to reduce the size of the syllabus. Better still, if you want to reduce content then do not try to cut-down but start with a blank sheet and only include material for which a strong argument can be made.

2) A change from didactic to reflective teaching

Since this requires reflection on the students' part, for which there is no time in the formal lecture, the lecture needs to be supplemented by resource materials. Alternatively, it could be interrupted by brief interactive periods (see Section 4).

3) Changing the structure

Students should be helped with their taking of lecture notes through the way that the lecture is presented using signposts and frames (see Table 2).

3.3 Student participation

INTERACTIVE LECTURES: GOOD PRACTICE
Cross out the activities you already employ

Split the class into small groups
Get students to sit at the front of the lecture theatre
Establish ground rules
Set clear objectives at the beginning of the lecture
Give students the chance to discuss the objectives
Feed tutorials into lectures and vice-versa
Stop talking
Have a break or time-out
Take written questions from students
Walk around and target the whole class
Use breaks to encourage questions from students
Identify individuals or groups to answer questions
Use humour, when relevant
Use appropriate body language
Write down student's comments and questions as they are received
Be aware of shy students
Allow students time to reflect
Use professionally produced overhead transparencies (OHTs) and slides
Use the board or overhead projectors (OHP) to draw diagrams
Check the availability and working of equipment
Role playing
Teaching not lecturing
Encourage students to think critically
Encourage students to prepare for lectures
Use handouts to summarise lectures
Use incomplete handouts for students to complete
Use appropriate teaching techniques for size of group
Encourage students to interact with each other

INTERACTIVE LECTURES: BAD PRACTICE
Cross out the activities that don't apply to you

Concentrate a two hour class into one hour
Patronise or insult students
Using sexist or racist language
Give students the impression you know everything
Give students the impression they know too little
Hiding behind equipment
Poor preparation
Waffling
Ask questions and then answer them yourself
Asking questions and not allowing time for an answer
Telling jokes that are not funny
Including too many anecdotes and asides
Antagonising students
Giving students very detailed handouts so they switch off
Underestimating student's abilities
Using OHPs for too long a time
Using OHPs for too short a time
Using tired data and information
Students only interact with you
Concerned with syllabus i.e. content rather than student learning

We have already stated the case for greater student participation in lectures as a means of facilitating deep learning. Active learning is seen as highly desirable as it promotes student thinking and participation, uses the time available more effectively given the 20 minute attention span of the average student and enhances deep rather than surface learning. There are two challenges for you to overcome: one is that the traditional lecturing style of the formal presentation discourages active learning; the second challenge is that with increasing class sizes it can be difficult for you to engage in audience participation. There are numerous suggestions in the literature for the ways in which students can participate in lectures, from texts such as Gibbs (1981) and Gibbs & Jenkins (1992), to the pages of the *Journal of Geography in Higher Education.*

Some ideas of how to work with students in small groups during lectures (from Brown, 1997)

Brainstorming	Students call out or write down ideas, words, names they associate with a topic
Pyramiding	Start with a simple often individual task and then get students to work in pairs and then groups as the tasks become progressively more difficult
Fishbowls	Arrange a group, perhaps the most domineering group first, in the centre of the room. Set them a task and require the rest of the class to act as observers
Crossovers	Mix the compositions of the groups after each task, perhaps by giving all students a number or a letter label
Buzz groups	Give groups small timed tasks that require them to discuss and then feedback their deliberations

Anderson (1994) provides even more suggestions, including the use of demonstrations, problem solving (such as case studies and role playing) and feedback (lecture review and topic synthesis).

Focused discussion classes

McBroom & Reed (1994) replaced traditional lectures by class discussion, short reading assignments, and daily testing for a course in social psychology.

Students are assigned daily readings of eight to ten pages that are easily mastered. The material is tested three times a week, which helps motivation, using 'short answer' and 'fill in the missing word' questions. If students do well (get an A) on the Friday test they are excused the following two tests. Initially students were concerned with trying to memorize the whole text but they soon became used to the system with 50 to 90% obtaining A grades on the tests. As tests were short there was much time left for discussion. This was facilitated by each student being required to prepare one question based on that day's reading. Sometimes the questions were merely factual but could provide opportunities for greater reflection, even going beyond the ideas and concepts contained in the immediate text.

(cont.)

For evaluation the course was taught by lecture one semester and then by class discussion. Comparison of student examination performance revealed that the number of incorrect answers was higher for those who had attended the lecture course. The authors concluded that focused discussion is superior to the traditional lecture format.

Using group projects and class discussions

D'Allura (1991) taught a ten week environmental geology course using a range of interactive methods. The course was introductory and attracted a diverse group of students ranging from 25 to 45 in number. In order to accommodate the class-based discussions, lectures were reduced to statements on key points, clarification and illustration. The methods employed included:

Personal experiences

Students were asked to relate their personal experiences to matters under discussion. It was found that class interest was heightened by listening to the actual experiences of the impacts of an environmental hazard such as an earthquake or a tropical storm.

Debates and discussions

Care needs to be exercised in the selection of controversial topics especially contemporary issues which can evoke strong emotional responses. Students were asked to back up their statements with supporting evidence and/or the class was divided into groups representing different points of view.

Group Projects

After the subject was covered during a lecture:

- additional materials were handed out;
- problem to be tackled was specified;
- students were separated into groups of 4 to 6 people;
- group discussions were held;
- class discussion were held, with communal sharing of knowledge;
- group answers were collected and graded.

The disadvantages of this approach included a possible increase in work load for the lecturer both devising projects, holding discussions and the marking of group projects. Students may not be well prepared and in large classes only a few groups may be able to present their work. The author feels that these problems are outweighed by significant educational gains. Students are more attentive and attendance is better. Students are enthusiastic about the group projects and perceive they have gained greater understanding and are less daunted by the scientific basis of the course.

If you are still looking for more examples there are various student participation exercises listed by Gibbs & Jenkins (1992) while Horton Smith (1996) provides a large range from his personal experiences including the trial, the quiz and the Oprah/Geraldo show. Horton Smith advises that one of the most valuable interactive techniques is to learn and use students' names.

There is then no lack of advice, but you may encounter teaching rooms with fixed seating where the idea of arranging large numbers of students into groups and even encouraging them to move around seems an impossibility. We will present the use of buzz groups as a case study Section 4 where this problem is addressed.

Making your lectures more interactive: from the following list mark off which the activities are found in the your lectures and then which of these could be taught with student participation:

Activities	In my lecture	Could be used interactively
Outline structure		
Explanation of key principles		
Demonstration of proof		
Demonstration (non-verbal)		
Hypothesis		
Problem solving		
Narration of events		
Causal explanation		
Comparison of theories		
Evaluation of theories		
Analysis of case studies		
Presentation of evidence		
Commentary on research findings		
Summarizing main findings		
Reviewing past materials		

3.3.1 Questions and answers in lectures

Merely asking a question will not necessarily initiate a discussion. There are many reasons why students do not respond such as:

- They did not understand the question, (due to lack of their understanding or the question was unclear).

- You did not give enough thinking time.

- Peer pressure and being afraid of giving the wrong answer.

- Students have forgotten the question while discussing the answer.

Here are some suggestions to overcome these difficulties:

- Count up to 20 (silently) before giving up.

- Write the question down on a OHT or give out as a handout.

- Suggest students make notes before asking for any replies.

- Suggest students discuss their answer first with their neighbours.

- Provide multiple choice answers.

- Ask open not closed questions (see below).

- Use Buzz groups.

We will discuss the last two here and in Section 4.

Many questions are closed, requiring a yes or no answer and should be avoided. For example, 'Is geography a science?' is unlikely to elicit more then a shake or nod of the head from a first year class. Open questions, such as 'What is the difference between studying for a BA and a BSc geography degree?', are better as they require a fuller response.

In your recent classes, can you recall any questions either you or a student asked that stimulated class discussion? Could these be used in future classes?

Using a questionnaire to encourage questions and student participation in lectures

Teaching air pollution in London

Asking 'what are the problems with London's atmosphere?' or 'what should be done to tackle urban air pollution in London?' to a first year group invokes little response. Instead an anonymous questionnaire is passed around at the start of the course. Questions are asked on students' views on the nature, causes and solutions of London's pollution problem. As this is completed at the start of the course answers are not influenced by the lecturer. At a latter stage the answers are summarised and returned to the students and then used to initiate discussion of road pricing, traffic regulation and other measures. The class size is 120, students from a variety of non-geography courses attend including overseas students but there is rarely a problem initiating discussion or eliciting answers to the questions posed at the beginning of the course.

Some examples of types of question and responses:

How does London's air compare with other British cities?

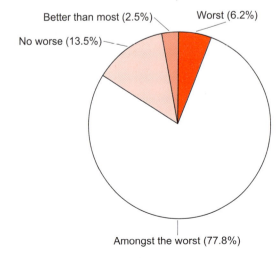

Better than most (2.5%)
Worst (6.2%)
No worse (13.5%)
Amongst the worst (77.8%)

These answers help to promote a discussion on the meaning of 'worst', the relative pollution of different British cities and sources of information.

List five air pollutants that are the most harmful to humans

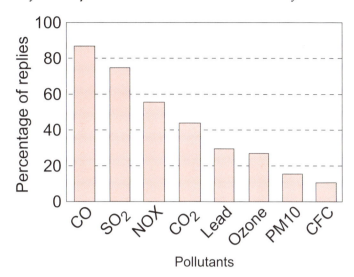

It is interesting to note that PM10 (particulate matter) receives very little attention despite recent reports linking this to asthma. CO and CO_2 are noted, possibly through confusion with greenhouse gases and SO_2 is listed although in London concentrations have fallen over the last 40 years. It is then possible to use these replies at the end of the course to question whether student's perceptions have been changed.

Should cars be banned in London?

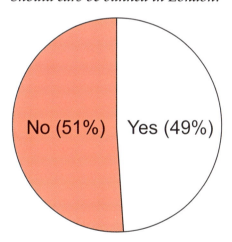

There is clearly a difference of opinion amongst students on how vehicle emissions can be reduced. Additional questions consider road pricing and student's own use of vehicles while studying at London.

There has been no formal evaluation of the use of the survey in the course but evaluation of the module on air pollution received favourable student evaluations:

Very Poor	Poor	OK	Good	Very Good
0 %	4 %	4 %	45 %	47 %

For further details contact Clive Agnew (cagnew@geography.ucl.ac.uk)

Having encouraged a response to your question, even if there is an inadequate or incorrect response, do you then ask another and then more driving towards a final, pre-determined destination? The answer is emphatically **NO**. This has been called the Alpine style, with a steady barrage of questions leading to the summit. It is likely to result in resentment and frustration as answers that do not fall on the predetermined path are ignored or deemed wrong. Student response is likely to be,

> *'If that was what he/she wanted us to say why waste our time asking questions and not just say so in the first place?'*

If you are going to ask questions then it is essential that you allow the answers to provide some direction for the next step; in this way the audience truly feels that they are participating and that answering the questions is worthwhile. It is difficult when none of the students feel able, or are willing, to answer the question. Providing possible answers, multiple choice or some variant such as a number of images and asking for a show of hands will enable the audience to participate. Subsequent discussion can then feed off the range of answers.

What should I do if a student's answer is wrong?

- Avoid questions for which there is a 'right' answer, e.g. questions of fact.

- If asking for a factual response is unavoidable and you get a wrong answer then go back to the class and ask 'how many agree?'

For example, do not ask,

> *'how many people in the world lack access to safe drinking water?'*

but present the students with a number of options by asking,

> *'which of the following do you agree with?'*

In 1990 1.2 million people lacked access to clean drinking water

In 1990 1,200 million people lacked access to clean drinking water

In 1990 1,200 billion people lacked access to clean drinking water

or better still ask,

> *'how can you find out how many people in the world lack access to clean drinking water?*

3.3.2   Questions posed by students

Are you prepared to respond to questions during your lecture?

If you are a lecturer who does not like interruptions, such as questions from students, reflect upon why this is so. There may be good reasons, such as the need to convey a substantial amount of factual material such as instructions for use of computing software or safety instructions for fieldwork. Questions may only be of interest to the student doing the asking and not the rest of the class. But you should allow for the possibility that disliking questions from students is due to your lack of confidence. You may be concerned that by allowing student participation it will then be difficult for you to regain 'control' of the lecture. Even if you decide to discourage questions from students there is one exception you should make, that is when you have made a mistake and a student is asking for clarification. It is much better to rectify a mistake immediately.

> • If you do not feel able to allow questions during your presentation, are there times when you will allow students to ask questions?
>
> • Students need time to think before they ask a question in a lecture. Do you allow students time for reflection and questions during your lectures?

You need to establish the ground rules for student questions and here are some suggestions:

- Explain what is expected: no shouting out, structuring the lecture into a series of parts at the end of which time is left for questions, asking questions that are relevant i.e. not wasting time.

- After a question has been asked by a student, repeat it back to the audience, in your own words to ensure all have heard and understand the question.

- The manner in which you respond to the student will determine to a great extent whether or not other questions are posed and whether or not students use this to waste time. Do not disparage the questioner. Even if you feel the question is ill conceived or shows misunderstanding you can use the opportunity to encourage the rest of the audience to participate through asking 'What do you think?'.

- Too many questions can destroy the coherence of the lecture so it is best to agree times in the class when questions are appropriate.

- A problem can arise with the mischievous question that is posed to waste time. In such a situation other students will feel resentment that their time is being wasted. If you feel this is happening suggest that the question needs more time to answer than the lecture allows and arrange to meet with the individual immediately after the class. At the end of the lecture leave sufficient time to remind the class about the rules for questions. Do not confront the questioner but spend some time to find out whether there is an attempt to be mischievous or it may be genuine confusion on the part of the student. In either case it will save time and effort to deal with the situation immediately.

- If you do not know the answer to a question state you will find out and respond at the next opportunity. Ensure that you do so, explaining how you obtained the information so that students can help themselves should a similar question arise later.

- If a question is posed such that it is based upon a student's own background, for example, a mature or overseas student, it may be that others will resent such an interruption. When responding and repeating the question to the audience, phrase it in such a way as to invite them to think about the implications from their own experiences.

- Your time is valuable. It is reasonable for you to expect students to have read required reading or to have tried to answer the questions themselves. Simply listening to the lecturer give the correct answer encourages laziness and passive learning. You may then decide not to answer the question but to suggest sources of information. In this case do try to remember to follow up the issues raised by the student's question in the next class. Such a response may be appropriate with a final year group but it does not help a first year student, for example, to be sent looking for a specialised journal that is not in the library.

3.4   Note taking by students

Although Gibbs (1981) reviews a range of studies that lead one to suggest that student note taking can lead to poorer learning, it is generally agreed that note taking enhances students' short term recall and can aid revision. Isaacs' (1994) questionnaire sent to 100 academic staff in Australia revealed that almost all those replying wanted students to take effective notes. This is not surprising given that a correlation can be shown between the contents of students' notes and their test scores. The most important functions served by note taking were noted as being (p.207):

- to provide students with a basis for further study;
- to help students remember what was taught;
- to provide a basis for revision;
- to allow students to see the structure of the subject.

Some of the Australian academics however rejected note taking as being useful on the grounds that:

- most of the lecture material is in adequate text books;
- detailed handouts are provided;
- notes are provided in a course handbook;
- they wanted students to listen, think, ask questions and to understand.

Of course merely providing lecture material elsewhere is of little use unless some means is employed to encourage students to access this information and make it their own. A very effective way to give students thinking time is to supplement the lecture by carefully constructed reading material or to give students access to an audio or video-recording of the lecture (audio is usually adequate and much cheaper as well as less intrusive). During play-back students can then stop and reflect at points where they find it appropriate and supplement the lecture by additional reading. In that way they improve both their understanding and their notes. Such recording is however largely pointless, if a lecture consists merely of the authoritative presentation of material; it must include passages specifically designed to stimulate reflection. It also encourages students to rely on the lecture when better use of out-of-class time is to read around the subject.

How do you check that your students have understood your lecture? Are all these formal assessments? How might you ensure that students are taking effective notes?

Do students take good notes in your lectures?

Hartley & Davies (1978) evaluated the effectiveness of note taking based upon the amount of information retained by students after a period of time. There was a moderate correlation between note content and test results but further analysis under more natural lecture conditions revealed that the quantity of notes varied considerably. Elton & Laurillard (1979) suggest this variation is introduced by the student themselves, their interaction with the lecture material and the method of instruction. They further point to the difference between the quantity of notes taken and their quality which are not mutually dependent. Hence it is very difficult to evaluate, or promote, 'good note taking'.

What constitutes a good set of notes is open to debate, whilst note taking is as much about student motivation as a lecturer's presentation. Here we are merely making suggestions of ways in which you might check what is being recorded by students in your lectures in order to learn what they have perceived as being significant and worth thinking about.

- *Look at students' notes during the class*

This helps you to rectify any errors and omissions immediately but it can be distracting for students and difficult in large classes. Students who have made mistakes can be embarrassed so some tact is required.

- *Ask questions*

See the discussion on questions above. This approach may lead to an emphasis upon factual content rather than a student's understanding.

- *Read students' notes*

Best done in a tutorial or other small group discussion when you can compare what you thought was important with what the students noted as being important.

Helping students to take notes:

Students coming to HE are often ill-prepared for lectures and their note taking skills are weak. At UCL an intensive first year induction programme is run with the aims of:

1. helping the transition between small class (school) learning and first year lectures;
2. familiarising students with basic IT skills and facilities;
3. making the department more welcoming.

(cont.)

The topics covered in the induction programme include:

The department	Who's who in the department
	Health and safety
	Dept. Facilities
	Careers
IT skills	Word Processing
	Libertas/Library
	Computing facilities
	Data presentation
Teaching	Lectures and note taking
	Writing an essay
	Maths revision & statistics
	References and bibliographies

In their evaluations of this induction training, two-fifths of students found the course good/very good, but one quarter were less then satisfied. Criticisms concerned the course being too intensive and held at a time when a lot of other University induction events were taking place, i.e. the first week of term. The lesson is that intensive induction programmes need reinforcement later. To address this the programme has been re-arranged from one intensive week to now cover part of the first week and then to continue through the first term supported by a weekly tutorial meeting as part of a new half unit course WAG (Writing and Analysis in Geography).

For further particulars on WAG contact Prof. Hugh Clout (hclout@geography.ucl.ac.uk) on the induction programme contact Dr. Ann Varley (avarley@geography.ucl.ac.uk).

You should help students with their lecture note taking through the way that the lecture is presented. Giving adequate time to copy from the overhead projector is the least effective of these, since students should always have access to such material. What is far more important is to structure a lecture carefully and to reveal the structure to students. Most lectures consist of several sections and it is absolutely essential to indicate to students when one section ends and the next one starts. In part this can be done through putting section headings on the OHP, but much more important is the old advice of

'say what you are going to say, say it, say what you have said',

applied to each section and in addition, the same applies to the lecture as a whole. If there is apparently not enough time to do this, then cut out material, but never cut out the first and third of the 'says'.

Lecturing will be more effective if the notes taken by students are useful and one way to do this is to reduce the quantity students have to write, for example, by issuing summary handouts and/or reducing content. The more structured and coherent a lecture, the easier it will be for students to produce organised notes. But is this effective learning or merely regurgitation? Perhaps the most important point is to ensure students understand the difference between **making notes** rather than simply **taking notes**, the latter being a passive copying down of information, the former involving synthesis and thought. This can be more effective if students are prepared for the lecture, i.e. through reading before-hand and if time is provided for reflection.

Tips you can give to students for effective note taking include:

- Write notes in your own words. This requires students to think about what has been said rather than straight copying.

- Being economical with paper may lead to confusion. Leaving lots of spaces enables students to add points raised later in the lecture and to cross-reference such information. Spider diagrams where notes start in the centre of the page rather than the top can aid this process.

- Advise against uniformity in writing down notes. Use different colours for facts, issues, quotes, sources. Underlining, using boxes and centring will help later recall of key points.

- Encourage the annotation of handouts: students can switch off otherwise feeling they have all the information they need.

- Suggest students continually examine their notes, distilling them into summaries, following up case studies, thinking about cross referencing different lectures.

- Advise against keeping all the notes in one file: if this is lost the result could be disastrous.

How lecturers can help students to take more effective notes:

- Indicate at the end of each lecture the material that will be covered next.

- Provide outline notes, particularly when mathematical solutions are being presented.

- Take time at the start of the lecture for students to note what they already know about the topic to be presented.

- Take time during the lecture to stop and reflect upon what has been covered.

- Suggest students write down three key points on the topic and then they should discuss these with their neighbour for a few minutes.

Helping students to take notes in your lectures

It is important that a lecturer, whatever his/her purpose, helps with note taking. It is particularly important to:

- reveal the structure of the lecture to the students;
- make it explicit when ending a topic and starting a new one.

Think of it in terms of how you would write rather than speak it, when you have the help of chapter headings, paragraphs and different ways of putting the point on the paper. All this has now to be conveyed by the spoken word with the additional handicap that the student cannot flip back the pages.

All this is right for the didactic lecture. It does not apply to the inspirational lecture for which note taking is inappropriate. But do you then expect students to remember what you have said? Students should come out of a lecture with a glow and want to learn, not a sheaf of notes.

(cont.)

There is an old story of the Provost of King's College Cambridge being approached by a student to borrow his notes, as the student was unable to adequately take notes in what was clearly an inspirational lecture on Greek mythology. The Provost handed the student an envelope on which was written:

Zeus
Agamemnon
Zeus

3.5   Following up lectures

A lecture is a guide, it facilitates learning and does more than simply convey information. Students will need to follow up lectures with their own reading and related activities. Students can do this in five ways:

1. *Consulting material used in the lecture, overheads, lecture notes and other visual aids.*

   It is recommended that you make available copies of the handouts and overheads you use, this way students will be less concerned with writing and be more prepared to listen. Open access materials tend to disappear and therefore always keep a spare copy or arrange a course file to be signed out by the library. This involves extra work for yourself and the library staff so you may consider using the WWW for storing this information so that a 'paper-less' course can be adopted.

2. *Obtaining additional material, for example, through reading lists, course handbooks, off prints and World Wide Web (WWW) sites*

   Reading lists are common place but there are many pitfalls: is essential reading indicated? Have you identified introductory and specialised reading? Are the publications available in sufficient quantities? Have you discussed the reference list with the students and indicated what you expect them to read, how many sources you expect them to consult? Again the WWW is useful.

3. *Listen to an audio or video tape of the lecture*

   See Section 3.6 and Section 4 where the use of lecture recordings is discussed. First year lecture courses at the University of Surrey were audio recorded and four copies were deposited in the library for one week, after which they were used again for another lecture (so as to discourage students who wanted to listen to all of them during revision). The audio tapes were used by approximately one third of students. The scheme was not evaluated, but it was incredibly cheap and clearly served a purpose.

4. *Consulting the teachers: office hours*

   Office Hours are increasingly important for allowing students to ask questions of staff following upon a lecture. Some staff still operate an open door policy, especially for emergency, pastoral or tutorial matters. However, posting hours when staff will be in their office and available for consultation helps to avoid unnecessary interruptions at other times. It helps students to plan their time more effectively and encourages them to think about the question before seeing a member of staff. An important issue is how many hours and when in the day? There is no universal answer as it depends upon other forms of contact, such as tutorials and problem classes and the amount of choice allowed in the timetable. For example, in the Department of Geography at UCL, in addition to tutorial meetings and seminars, each member of staff announces three hours a week of 'office hours' arranged at any time. If all three hours clash with a student's classes then they will arrange to see the member of staff privately. Email is now being used by students rather than office hours for consultation but there are some pitfalls.

Using email for staff:student consultation

The advantages of communicating with students by email include savings in time and photocopying. There are however a number of issues:

1. Email or Office hours?

A student may prefer to email you than to come to your office. This can be mutually convenient as you can decide when to respond i.e. better time management for all. But does it take more time to read and respond to an email than to talk to a student directly during your office hours. Are you now spending more time on consultation i.e. office hours plus email?

2. Length of message

You want to avoid students sending you lengthy emails. Complicated matters are best dealt with in person. Emails from students should then be concise questions and not matters for discussion.

3. Subject referenced

Are you prepared to receive emails on any topic or only related to your course(s) and/or tutees? Whatever your decision this should be made clear to students. It is also useful to agree that all emails from students should include the course reference in the title. You might also insist than when a student responds to one of your emails they retain your message title so that you can identify the nature of the message at a glance.

4. Response time

How rapidly can students expect you to answer their queries?

5. *Setting students tasks and supplemental instruction*

Post-lecture activities include practical and problem classes perhaps with computer-assisted learning, group discussions (formal or informal) and writing assignments to list but a few. These activities can be enhanced with supplemental instruction, using senior undergraduate or postgraduate students to assist in classes where there is a difficulty such as lack of background or technical skills. McCarthy *et al.* (1997) examined the use of supplemental instruction in South Africa and concluded that there were discernible improvements in performance for those students who did attend such classes particularly for students from disadvantaged backgrounds. There are problems with this approach, not least finding suitable students to act as instructors and an appropriate method to reward them. When second year 'facilitators' were used at UCL to help first year undergraduate students who lacked numerical and computing skills the initial enthusiasm of the instructors waned as coursework and other assignments became due.

Using writing assignments to augment lectures

Macdonald & Conrad (1992) use writing assignments to support the teaching of an introductory geology course. Writing is a skill and lecture courses that emphasise content ignore the need to refine this ability in HE. The authors argue that this approach involves students more directly in learning through a process of self discovery which leads to greater enthusiasm and better understanding, but caution that the assignments have to be carefully designed. They make four suggestions:

Effective design

- Provide frequent and varied types of assignments ranging from in-class work, papers of different lengths, and reviews to problem solving.

- Demonstrate writing as a process by explaining what you do.

- Link assignments to course objectives.

- Set clear guidelines which specify your expectations of the work such as whether or not it should provide critical insights.

Teach writing skills

This can be aided by discussion of the criteria to be used in the assessment of the written work and the use of sample essays to illustrate good and poor practice (see Writing and Analysis in Geography example in Section 3.4, p.29)

Peer review

The exchange and marking of papers between students can greatly aid students' understanding of what is expected, but peer reviewers need to make constructive criticisms and to focus on structure and content.

Formal and informal assignments

Informal assignments include questions and answers, in-class writing, writing notebooks, response papers and memos.

Formal assignments include abstracts, laboratory reports, letters, short papers, research reports and research papers.

Problems associated with this strategy are: the reduction in lecture content to make way for the writing assignments; resource constraints if similar materials are to be used by a large number of students; the time required to evaluate the written work, especially in large classes. Although the authors have not assessed the long term benefits of using writing assignments they find students respond positively and believe the approach encourages students to think critically and read more carefully.

3.6   Use of visual aids

3.6.1   The use of overheads

Overheads have become an essential tool for many lecturers. They are useful because they can be prepared beforehand. Here is a compilation of points for you to take into account (based on Race & Brown, 1994, p.52):

- Ensure that each transparency is visible from the back of the room. Using a 24 point type size is recommended.

- Include the source of any data, figure, quote etc.

- Check that the transparency will fit the projector.

- Avoid using transparencies with lots of words or data: if necessary produce these as a student handout.

- Make your overheads available, perhaps by placing copies in the library or on the WWW.

- Make sure the overheads are professional in appearance. If they look scrappy then the audience is less likely to trust their message and will have less confidence in the lecturer.

- Take time to talk through the overhead. If it is a graph or a table of data explain the labels on the axes or columns, the patterns evident and the source. Students who are learning will need more time to assimilate this information compared with those who are merely looking or copying.

- Annotate transparencies during the lecture to draw out key points.

- 'Progressive reveal' (i.e. masking the transparency and slowly drawing back the mask to show the information) gets the audience to concentrate on one point at a time but can lead to frustration if used too often or if the full transparency is only displayed for a short time.

- Progressive layering of several transparencies to illustrate increasing complexity can be a useful technique but is difficult for students to take notes. To avoid frustration provide the final complex image as a handout.

- Use colour to add variety but note that orange and yellow do not project well while red may not be clearly visible in a large lecture theatre and there may be students who are colour blind.

- Transparencies are flimsy. If you are going to use the transparency repeatedly try to secure it in a card board or plastic sleeve.

- Ensure you understand the different types of transparency, i.e. those that can be placed in a photocopier or laser printer without melting.

- Never put something onto an OHP that does not relate to what you are saying, simply as a way of covering more material.

The copyright of any material used in slide form, acetate or transparency needs consideration. This will be covered in England by the Slide Collection Licensing Scheme negotiated with Design and Artists Copyright Society, London. However, this does not cover Ordnance Survey maps, film stills or advertisements.

3.6.2   The use of video, interactive multimedia and IT

(See also the GDN Guide by Shepherd, 1998)

Students today are familiar with videos and there is no longer any impact in using this technology apart from the change in presenter or pace. It is then necessary for you to consider carefully the usefulness of this approach. It is possible for students to be directly involved in the making of videos (see below) but here we are concentrating on the use of 'moving images' in the lecture setting. This no longer solely concerns the use of video cassettes or 16mm film but also encompasses the use of video sequencing using presentation software such as PowerPoint or earth observation image processing facilities and computer model simulations (see the CTI WWW pages http://www.geog.le.ac.uk/cti/tltp and Bishop *et al.*, 1995 who provide details of WWW sites that include video animation and satellite images). The essential difference between video and other imagery used in lectures (for example, slides and OHPs), is that video deals with a moving image and this offers many advantages by bringing to the lecture immediacy and life. Experts can be listened to, a hurricane is 'experienced', the past is visited. Using a WWW browser the distinction between static and moving imagery is far less clear and integration is much easier, but there are problems over access to the WWW during a lecture.

Sources of film and video

The WWW has greatly added to the range of visual material that can be obtained and used in lectures but it can take time finding appropriate sites. Some are listed at the end of this Guide but you may like to look at the following sites which provide information on a range of products although not always geographical.

British universities film and video council

http://www.bufvc.ac.uk/

Anglia multimedia

http://www.anglia.co.uk/education/

BBC education

http://www.bbc.co.uk/education/

Sumner (1984) argues that in geographical topics such as meteorology which deal with a dynamic system operating at a variety of scales, time-lapse video can help students understanding of the nature of change and the processes involved. The static pictures presented in books or by slides do not encapsulate the dynamics of the atmospheric system. At a macro level climatologists are trying to explain changes that take place over thousands of kilometres that video sequencing of satellite images taken over several days can summarise. Time-lapse studies of clouds or the movements of fronts can lead to greater understanding of their formation and development. Gunter (1991) found interactive presentations of three dimensional images greatly enhanced the teaching of earth science.

Some tips for you on the use of videos:

- Attention spans are limited, it is better to use a number of clips than one long piece.

- Do not talk while the video is running, unless it is a silent video.

- Students need to know what is expected of them, they are used to watching television without taking notes. Consider the use of a handout with questions that they answer during or after the video.

- If students have discussed the topic before seeing the video they are more likely to make use of its content.

- Can the video be seen and heard by all of the audience?

- If students want to refer back to the video in an essay agree an appropriate form of citation and provide the necessary details.

- A summary OHP transparency at the end is a useful way of ensuring students will remember the content and issues.

- It is difficult with a large group to answer questions of detail or to respond to requests for replaying particular sections. Make a copy of the video available to students so that they can go through it at their own pace.

- As a rule of thumb, the distance between the audience and monitor should not be greater then ten times the screen diagonal length for video, and only six times for data. This means that monitors are not suitable for teaching groups of students larger than 20 to 30 unless screens are suspended from the ceiling. In larger class rooms data-video projectors are required

There are, however, a number of problems associated with video which include the use of special editing facilities to produce clips, the greater time taken than to prepare OHP transparencies 'in-house', copyright restrictions, and obtaining (booking) suitable facilities.

- Remember that video is a powerful medium and students need to develop a critical approach to the messages and information conveyed.

- Why are you using a video — is it really necessary? Explain to the students why these images are so useful.

Murphy's law and the use of television monitors

Flowerdew & Lovett (1992) reviewed their own experiences in using computers in geography lectures referring to 'Murphy's Law', that anything that can go wrong will go wrong. Economies of scale dictate that it is more efficient to place data-video projectors or computer monitors in large lecture theatres. They recount the use of a 300 seat lecture theatre with monitors arrayed along the sides, which when occupied by a class of only 65 students produced a distribution of seats being occupied around the periphery with none in the centre — a difficult arrangement for teacher and students. The class was often looking to the monitors at the sides of the room and not at the lecturer who was standing at the front. For security reasons computers could not be left in lecture theatres so each class had to be delayed while the connections were made with the predictable problems of wrong cables, hardware and software faults. If it could go wrong it did.

McCartney *et al.* (1992) describe how they used television broadcasts to teach introductory physical geology and laboratory courses at the University of Maine. Two remote classrooms were created with individual microphones for the students to aid interaction with the lecturer while additional sites had instructors in the room to assist with questions. Computer networks now permit the transmission and display of video, audio and graphics across much larger distances. Distance learning can make use of this technology and video conferencing through the Super JANET and ISDN data networks is now possible across the world. A major concern with this approach is that it continues to promote the traditional style of lecture, that is the broadcast of a talk to a large group of students who remain passive. Wright & Cordeaux (1996) argue that recent developments in desktop computers now permits the use of desktop video-conferencing where students and lecturer can be more interactive. Kies *et al.* (1997) have evaluated the possibilities of using desktop systems for video-conferencing where problems centre on the degradation of displayed images due to bandwidth restrictions. They conclude that student performance, as measured by a factual test on the information displayed using different image resolutions, frame rates and audio, video and audio with video, does not suffer from reduced video quality but noted some difficulties in using desktop video conferencing across the Internet.

Lecturing by video conferencing: Live-net

There are many examples when the broadcasting of lectures can be useful such as a University occupying split sites, students working from home or a lecture theatre having insufficient capacity (see case studies in Section 4 for more detailed discussion of distance learning by televised lectures).

Live-net2 is concerned with the experiences at the University College London. It is based upon the Super JANET and ISDN data networks linking sites at Cardiff, Cambridge, Manchester and Edinburgh to list but a few. Live-net2 studios are equipped with facilities for broadcasting VHS video playout, computer screen display and other visual media.

The advantages of the method include:

- Large numbers of students can be reached.

- Use of an overhead video camera can enhance the presentation, particularly demonstrations.

- Efficient use of teacher's time.

Evaluation: The system was used to broadcast a geography lecture course to colleges across London on Global Environmental Change organised by Royal Holloway & Bedford New College. The technology presented no serious problems but student exam performance was poor and the course only ran for three years. Today more use is made for medical teaching than for geography students. Medical students need to see demonstrations of techniques and there are savings to be made by broadcasting this to a wide audience that involves the major British teaching hospitals. Geography teaching does not normally require demonstrations nor are similar technical skills currently emphasised.

The experiences gained using Live-net include:

1.   The lecturer presentation needs to be changed. The lecturer is not free to wonder about and less detail can be included on visual aids.

2. To enable students to respond during the lecture, rooms have to be equipped with microphones, and therefore special teaching accommodation may have to be used although desktop video conferencing may change this requirement.

For further details contact the Live-net office 0171 580 9872 or http://av.avc.ucl.ac.uk/livenet

Use of film to teach geography

Source: Jenkins & Youngs (1983). Geographical education and film: an experimental course.

Context: When this study was first presented there had been little use of film and even less of video to teach geography. That has since changed, see the account by Gold *et al.* (1996) on the 'Grapes of Wrath', but the lessons to be learned remain the same. In essence students and teachers need to be more aware of the bias present in film, the power of such imagery and to develop their critical assessment of this media (see Rose, 1996).

Aims: The aims of this study are to develop students' critical evaluation of film, to understand the choices made by the film maker, to see the links between the film and geography as a discipline and to gain some competence in film making.

Method: Several films were observed but two were used for detailed examination: a portrayal of Ludlow, a market town, and a film on the Fens entitled 'A Sense of Place'. Teaching ran over 10 weeks and included discussions of the films, assigned reading and making a video.

Evaluation: Students found it difficult to find the links between the films and geography but felt they had gained in their critical evaluation of film content. They were enthusiastic about the course but many were dissatisfied with their own video production and it would appear that this element of the course works best if students are already familiar with film making.

3.7   Problems with students

Guides to teaching students in HE tend to assume all students are well behaved, attentive and reasonably enthusiastic. We need to recognise however that students coming into HE from school have been taught in relatively small classes with probably a seminar format. The large lecture theatre is then an alien environment and some may have difficulty adjusting to these new surroundings. As class size increases the lecturer is increasingly removed from part of the audience. The greater the number in the audience, the more the lecture becomes a 'performance'. Eye contact is difficult and background chatter from the rear of the room may become disconcerting or even disruptive. These are also actors' problems, since they have difficulty in seeing their audience in a darkened auditorium and rarely are in a position to discipline a noisy audience.

How do you cope with a noisy audience?

Walk out in a huff, suggesting that they need only attend if they are going to listen?

Shout at the top of your voice or make some other loud sounds?

Elicit the help of colleagues to sit amongst the audience or come into the room?

Switch off the lights or even turn them on?

Talk with decreasing volume until the audience listens?

Most if not all of these will only bring a short term respite. But is background chatter a problem for you? You can gauge whether or not the pace and/or material is appropriate by the amount of glances and words passed between members of the audience. Background chatter, if not too loud, can be a valuable aid. Perhaps a more important point is that if you are encouraging student participation and discussion then should you be worried when this takes place? The answer of course depends upon the amount and timing of this chatter. A high level of noise will drown you out and prevent buzz groups or other methods of student participation from taking place effectively. There will be times when you want the audience to listen and pay attention to what you are saying, there are other times when you want them to discuss and talk amongst themselves. The issue is then how to establish appropriate behaviour for each of these activities and how to signal to students which mode of delivery the class is embarked upon. Here are some suggestions:

Agree rules of behaviour

At the outset establish the rules of acceptable behaviour and do not allow breaches to go unnoticed. Explain to the students why you want them to be quiet at certain times and how you will indicate when you do want them to discuss and participate orally. The problem with this advice is how to respond to 'breaches' of agreed behaviour. Often peer pressure can be used; talking to the students or their tutors can be useful but in very large classes this is often impractical. Certain types of behaviour such as reading a newspaper in class or shouting out should be dealt with through the appropriate disciplinary procedure; after all as the lecturer you have the responsibility to ensure that all students have the opportunity to learn from your class. Disruptive students should be asked to leave the class as they are preventing other students from learning. As a last resort you should be prepared to walk out of the room, this rarely happens and is therefore usually highly effective. You may refuse to teach the class until the cause of the disruption has been resolved.

Cause a distraction

If the class is noisy at the outset there are several techniques that can be employed. Dimming the lights may work but a more effective approach is to have some work tasks for students at the outset. This may require them to write down the three key points from your last lecture, you may put up an overhead with data or an interesting image and with some questions. Ask them to write down three things they know about XXX.

Explain your signals

The opportunities for student participation are usually clear as you will have set some task, organised buzz groups and so on. You then need some signal for this activity to cease. Agree this beforehand, whether it is the switching on and off of the overhead projector or simply the clapping of your hands. In this way the students will know when it is time to discuss and when it is time to listen.

3.8 Costs

There are real costs involved in the change from passive to active learning in lectures that need to be borne in mind.

Equipment: Facilities have to be commensurate with the purpose of the lecture and size of the classes. Monitors ranged along the sides of rooms can present difficulties and it is desirable that large teaching rooms are equipped with a PC and software such as PowerPoint and video display facilities.

Rooms: Although we have suggested you can encourage student participation in large sloping lecture theatres, the environment and possibilities are improved in rooms where the furniture can be moved. This may be too expensive or impractical for large classes.

Consumables: Many of the suggestions above include handouts and materials that are distributed to students during the lecture, i.e. they cannot be placed in a library for later consultation. There is obviously a cost incurred in their production.

Staff: The effectiveness of student participation in large classes can be enhanced by additional teaching staff such as teaching assistants but there are problems (see Goodland, 1997 and discussion in *Journal of Geography in Higher Education*, 1996, 20(1), pp.83-122), while using more staff in lectures could be seen as counter-productive by the institution.

Time: Perhaps the greatest obstacle to improving your lectures is the cost in your own time. It will take time to think about ways in which students can participate and to prepare the necessary materials. There are some savings to be made through a reduction in content and time spent formally addressing the audience. Nevertheless, you will have to recognise that it is necessary to set time aside for teaching preparation.

3.9 Preparing for the lecture

We have left consideration of preparation to the last as it depends upon the style of lecturing you have decided to adopt, the structure, format and the amount of student participation. Preparation will include you thinking about the context of the lecture, deciding upon the structure, writing an introduction, discussion and a conclusion, preparing handouts and visual aids, and finally rehearsing.

Lecture preparation

Here are some questions for you to answer while preparing for your lecture (from Hay, 1994)

Who is your audience?	(class size, background etc.)
Where are you speaking?	(controls, fixed seating, OHP and slides at same time?)
How long will you speak?	(rehearse your timing)
Why are you speaking?	(to educate, to entertain, to recruit?)
What is your subject?	(does subject match presentation?)
Have you done your research?	(collect up to date information)
Is there any dross?	(avoid padding with well known facts)

Brown & Atkins (1988) suggest that the essential skill in lecturing is adequate preparation not presentation. Realise that it will take much longer to prepare your lecture than expected. The most common mistakes are to include too much material, out-of-date material and figures/information that are too complex for the audience readily to understand. Cox (1994) makes the following suggestions:

Finalise your preparation

Read through your notes and look through your visual aids and handouts until you are thoroughly familiar with both the content and the purpose of the lecture. Pick out key points, concepts, information.

Mental preparation

You cannot think as quickly when lecturing as part of your attention is being used to monitor the class and responding to the audience. It can also be stressful so the more familiar you are with the material, your audience, and surroundings, the easier it will be to concentrate on the lecture delivery.

Accommodation

Check the room is suitable for the type of lecturing you have in mind. Can all the facilities be used at once or are there shared projection screens? Can the lights and ventilation be controlled from where you will be standing? Will you block the view of the board/screen for some members of the audience?

Timing of the class

9:00 am Monday and 4:00 pm Friday time slots are not the most popular, but in a busy programme may have to be used. Are the time slots allotted to you suitable for the type of class you will be teaching? For example, if Wednesday afternoons are kept free for sport will this mean half of your discussion groups disappearing on Wednesday mornings when there are away fixtures? Do you need to space lectures through the week to enable students to work between classes? Most lecturers of course do not have control over their timetable slots.

3.10 Summary

Having considered the mechanics of lecture preparation and delivery, and the range of approaches that can be employed, it should be clear to you that there are various options. There is no single best way to lecture but we have emphasised the importance of involving students. That is we have tried to promote deep learning in lectures. The rest of this Guide outlines a number of case studies.

Case studies

4.1   What is good teaching practice in lectures?

There are a range of views on what constitutes a good lecture. Hodgson (1984) noted that while staff emphasised content and the development of critical thinking, students rated presentation, clarity and enthusiasm for the subject. Brown & Atkins (1988) suggest that a polished lecture which provides a solution to a problem may be considered 'good' if conveying this information was the purpose, but if the student was expected to consider alternative solutions or to think critically about the solution offered then the lecture was ineffective. A good formal lecture generally involves:

- Clear objectives.

- An appropriate pace and level of delivery.

- The use of audio-visual aids.

- Enthusiasm and understanding of the topic.

- The raising of questions, i.e. the development of a critical approach.

- Clear and coherent structure.

Questions such as "Is the room adequate for the style of teaching?" "Do the audio visual facilities work?" "Do the results reflect students' abilities" or "Do the members of staff have the relevant background?" are reasonably easy to assess with a degree of objectivity. But many criteria such as "Did the students learn anything?" are not readily quantifiable and assessment involves a degree of subjectivity. Jenkins & Smith (1990, p.131) provide the following extract from an HMI visit which illustrates the types of judgement being made:

'Lectures contain up to date information and are sound in content. Audio visual aids are used at times and students are regularly provided with suitable hand-out material. The predominant style of lecture and seminar sessions is of transmission of information to students…. There is an emphasis on teaching but little verification if learning takes place. Few questions are asked by staff or students…the two sessions judged to be unsatisfactory or poor lacked variety in activities and suffered from a lack of clarity in purposes. Students in these classes were not encouraged to be involved and were often clearly bored or uninterested.'

TQA class observation notes

The recent 1994/95 HEFCE evaluation of geography teaching in England (HEFCE, 1995a, 1995b; Chalkley, 1996; Healey, 1997) was carried out by peer review with lecturing evaluated through Teaching Quality Assessors attending classes. The following aide-memoir was used to record the assessors impressions:

Room				
Length of Observation				
Students	Male	Female	Total	Composition
Subject				
Type of Class	Lecture/Seminar/Tutorial/Workshop/Practical/Other			
Teaching	Objectives, planning, content, methods, pace, use of examples			
Student Response				
Accommodation and Resources	Room layout, specialist equipment, use of visual aids			

Jenkins & Smith (1990, p.128) argue that

'Quality has a meaning only when related to functions such as fitness for purpose.'

but this is often ill-defined with respect to learning outcomes. Jenkins & Smith (1993) later list measures such as output (degree performance), value added, meeting student (customer) requirements as possible measures, but caution that staff and student classroom experiences remain important. With this in mind we have not carried out a quality assessment on each of the case studies presented below, but have tried to include approaches and techniques that have been observed to work with an evaluation of the possible difficulties and costs. The case studies are drawn from a variety of sources. They merely summarise and illustrate possibilities and you are strongly advised to examine the original sources in full. Some are presented as paired examples to provide some comparison of alternative approaches.

4.2   Case study: how do I encourage active learning?

Aims: To encourage active learning through student participation, particularly in large classes.

Source(s): This case study is based upon ideas from Charman & Fullerton (1995) and Jenkins (1992), but see also Brown (1997), Cryer & Elton (1992) and Gold *et al.* 1991. See also abstracts on the GDN WWW pages: http://www.chelt.ac.uk/gdn.

Context:

Gibbs & Jenkins (1984) argue that student learning in lectures will improve if each period of listening by students is no more than 20 minutes. It follows that breaks and activities need to be introduced. These need not be long pauses or non-activity but should be used to stimulate students into thinking about the new knowledge they have just acquired. The advantages of small group teaching have long been understood and have recently been re-emphasised with the growth of 'enterprise' in HE and the development of students' transferable skills. There is now recognition that small-group teaching techniques can be employed even in large classes, but that this does present particular challenges. The examples cited here, the use of buzz groups and class discussion, are but two examples of possible approaches that range from role playing simulations and debates through to public enquiries (see Anderson, 1994; Gibbs *et al.*, 1992 for a large number of ideas and suggestions).

Implementation:

Using buzz groups

Principles:

Buzz groups are named after the hubbub and chatter created by small groups of students engaged in discussions during the lecture. Buzz groups should not be used merely to revive attention, rather they should engage students in thought as well as discussion. This can be encouraged by asking for a minute's quiet individual reflection before the buzz commences.

The technique is to suggest some activity and then to divide the class into a series of small groups. This is easiest when students are already arranged around tables but even in a raked large lecture theatre it is possible.

Figure 5: *Arranging students into buzz groups in rooms with fixed seating (from Cryer & Elton, 1992)*

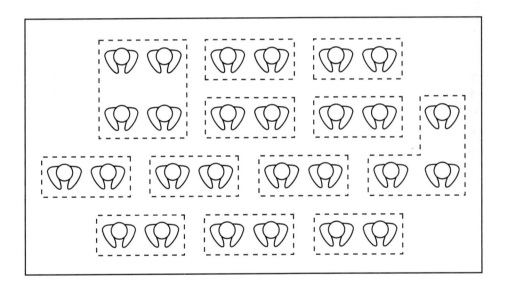

Much depends upon the type of activity required. This must be tailored to the students' ability and knowledge — it is no use presenting them with complex data or problems beyond their comprehension. Use an image, a brief piece of writing or video extract.

Students need to appreciate the purpose of using buzz groups. They need clear instructions on what to do, how long to do it and what is expected from them at the end. These instructions must not just be oral; they should be on the OHP.

Arranging groups:

Students will normally be sitting next to someone they know so there is usually no problem in getting them to talk to one another. Sometimes, especially at the start of the year, this may be more difficult. So begin with a simple task requiring students to introduce themselves to the person sitting on either side by stating which is their most and least favourite meal/drink/country/town, or simply their name and where they come from.

In Practice: (see Figure 6)

It may be that the discussion is so animated that it is difficult to stop the process. Agree beforehand the signal for discussion to cease possibly by turning on/off the lights.

In a large class there will not be time for all the groups to respond. Ask two or three volunteers for their thoughts and then ask for any other groups who would like to add anything new. Don't respond immediately to any of the reporting back, just make encouraging signals, a nod of the head, 'that's interesting, good points'. Any response, especially if it is critical, is best depersonalised and incorporated into the subsequent lecture. It is essential to time-limit the discussion and tell students when they have to complete the task, plus when there is one minute to go and whether they will have to report back.

Figure 6: *A class on Central Place Theory (from Gibbs & Jenkins, 1984; Jenkins, 1992)*

Stages	Time	Activity
1	5 mins	All: a revision OHP is displayed of previous work class has done on the topic while students enter and sit down.
2	9 mins	Lecturer: a revision talk on previous work
3	5 mins	Students: display the question "What aspects of C.P.T. can be used to analyse the number and location of shopping facilities in towns? Students discuss the question in small groups.
4	7 mins	Lecturer: short talk on set question with student questions
5	4 mins	Students: a new task is set employing data displayed on the screen
6	6 mins	Lecturer: summarises using feedback from some student groups
7	6 mins	Students: set a harder task interpreting and examining changes in locations of towns through time using C.P.T., working in pairs
8	1 min	Lecturer: answers part of the question leading to a more advanced issue
9	2 mins	Students: continue to work in groups
10	3 mins	Lecturer: complete the analysis in short lecture
11	1 min	Students: class is set an open question with insufficient time to answer
12	6 mins	Lecturer: reviews the material in the handout
13	2 mins	Students: write a brief summary of the lecture

Using interactive lectures

This approach was used to teach a geographical concepts course where problems had been encountered in the past when consideration was given to perspectives in physical geography. The traditional lecturing approach had not been successful and it was felt students needed more help to develop their understanding of theoretical ideas and their significance. The aim was promoting deep learning by increasing the amount of time students spent thinking. The approach employed was:

Handbook   Students were issued with notes at the start which identified key points covered in each lecture

Discussions   Students were asked to work in small groups at intervals within the lecture:

> 10 min Questions from students about previous lecture
>
> 10 min Introduction to class
>
> 10 min Case study explored by students in small groups
>
> 15 min Feedback from students
>
> 10 min Review by students individually and in groups.
>
> (you will note that the formal presentation by the lecturer is significantly reduced)

Question Box   Students who did not want to speak out in the classroom could ask questions by placing them in a question box placed by the door. These questions were answered at the beginning of the next lecture.

Comparison:

Both approaches used small-group discussion to promote active learning in a lecture room. The organisation of the time was broadly similar with the lecturer acting more as a facilitator as students were encouraged to investigate issues.

Evaluation:

Through the use of discussion groups it is believed that students develop their understanding of a topic, they develop communication skills, the attention span in a lecture is extended and deep learning is facilitated. There are constraints including suitable accommodation and the preparation of handouts and suitable tasks and activities. It may also be rather intimidating requiring students in a large lecture theatre to be organised into groups and then required to complete an assignment.

The use of buzz groups requires additional preparation time. The tasks have to be achievable, the questions answerable and set in the context of the lecture. It may be that images, data, multiple answers have to be prepared as handouts or an OHP transparency. The time required, however, is modest, perhaps a couple of hours including the planning. There is some gain in the time spent preparing for the lecture in that the lecture has to contain less material. There is a major gain for the students learning and their confidence to converse with colleagues and lecturers. It is also a very pleasant teaching experience to engage a large audience in conversation and to work together towards greater understanding of a topic. There is only a modest resource cost.

The evaluation of student responses carried out by Charman & Fullerton (1995) revealed a high degree of student satisfaction with only 3% feeling they preferred the more traditional lecture format. Although the typed notes were appreciated almost 60% had not read them before the lecture and only 19% ever asked a question in class. Overall it was felt that lectures had become more stimulating for students and more rewarding for teachers. It was noted however that the low audience participation needed further consideration and the lecturer needed to be able to move amongst the student discussion groups.

Summary:

These are not a new ideas but require confidence to be used with large classes. Doubts such as 'will the students do what I ask of them?' are bound to be felt. There is then much need to prepare the student audience properly and to persuade them of the benefits of this approach.

4.3   Case study: lecture reviews by students

Aims: To use topic synthesis (i.e. lecture review), at the start of a lecture as a student-centred learning approach in order to develop both critical thinking and communication skills.

Source: This case study is based upon the papers by Mossa (1995) and Newnham (1997).

Context:

It is good practice for lecturers to summarise their previous talks at the start of a fresh lecture. This places the current lecture in context and can remind students of key issues and points that need to be cross-referenced with the topic(s) about to be covered. It has been suggested in this Guide that this opening period can be used to stimulate student interest through discussion of either the students' own understanding of the material to be lectured upon or by reviewing some previous work or activity. This case study presents two strategies where this has been employed: one in the USA and the other in England. Newnham also notes that lecturers who employ the traditional didactic lecturing approach and feel uneasy about losing control may find the lecture review strategy more acceptable as it takes place at the start of the lecture and provides useful feedback.

Implementation:

The approach employed by Mossa (University of Florida) is presented first, followed by the method of Newnham (University of Plymouth).

Topic synthesis

Mossa was concerned that students would feel apprehensive about public speaking and time was taken to explain the objectives and what was expected. Guidelines are provided with advice for effective presentation. The presentations are organised around 10-minute talks at the start of each class by one or two students. The tasks for students are:

- Sign up for and attend one of the classes.

- Summarise the key points and concepts from the previous lecture.

- Demonstrate comprehension through discussion of which issues were most relevant.

Although this was not assessed formally the presentation was included in the 15% mark for class participation. This assessment included errors of omission, inaccuracies and overall content and coherence.

Lecture synthesis

Students were formed into groups of 4 to 8. One student from each group prepared a written review of the previous lecture on a single sheet of paper that is presented to the rest of the group at the start of the next class for a period of 3 minutes. A further 3 minutes is devoted to group discussion. The sheet, annotated with new points, is then given to the lecturer. The sequence is then:

- Student speaker presents lecture synthesis to rest of group.

- Group discussion of the synthesis.

- Summary sheet given to lecturer.

- Student reviewer for each group selected for the lecture to come.

- Lecturer comments upon reviews completed for previous lecture.

Comparison:

There are many similarities between the two case studies. Both had fairly small classes (20 to 30 students), formal assessment was not included and both limited the total time devoted to the feedback and discussion to a maximum of 10 minutes. There are differences in the logistics but not in the basic objectives. Newnham prefers students to discuss the work amongst themselves whereas Mossa opts for a more formal presentation by one or two students at the start of the lecture. Mossa stresses that students must not regurgitate but synthesise and evaluate, whilst this is not noted by Newnham. With Mossa the feedback by the lecturer takes place immediately after the student presentation whereas Newnham does this at the next class after the summary sheets have been completed.

Evaluation:

There are a number of potential benefits for students including:

- Developing presentational skills. Students need to become effective communicators, hence develop their oral and presentational skills. The strategy outlined above provides an opportunity for students to practise and hone these skills.

- Promoting deep learning. This approach requires students to think critically, to evaluate each lecture, hence promoting deep rather than surface learning.

- Effective note taking. A good presentation requires that a student pays full attention and takes accurate and effective notes.

- Variety. Having more than one main speaker adds variety to the lecture.

- The lecturer is better able to gauge how well students understood the previous lecture.

- The lecturer may obtain invaluable insights into the understanding of particular students and be able to identify any experiencing difficulties.

Problems did occur. In Mossa's case, some students talked for too long or tried to repeat the lecture rather than synthesise and evaluate. The approach by Mossa could be seen as a 'trial by fire': an audience of peers can be highly critical. The tasks require students to think critically, to take effective notes and to be able to present their ideas, but this could easily become a frightening and intimidating experience. A key issue is how are the students taught and helped to achieve these aims?

Newnham's approach encouraged discussion especially for students who were uncomfortable at presenting their views to a large audience. It would be problematic if applied to a large class both in terms if suitable accommodation and time if several groups of 4 to 8 are to meet, talk and then listen to a lecture. What works for large classes can work for smaller classes but not necessarily vice-versa.

When asked how this method could be improved, Mossa's students suggested that the 10 minute timing should be strictly enforced and that the summary be produced as a 1-page handout for circulation. Despite these reservations over public speaking the student evaluation of both of these exercises was highly favourable. In Mossa's case all students felt it helped their retention of material and 84% felt their comprehension had improved. All of Newnham's students felt the lecture review sheets were useful and 67% felt the process worked well and should be continued.

Summary:

Mossa's approach was based upon a mixed ability class with a combination of graduate and undergraduate students which may have added to the nervousness of the less experienced. The class size was small (20 students). Mossa suggests that beyond a class of 40 students (Newnham suggests 70) problems may be encountered, especially for students uncomfortable with public speaking and, of course, if everyone were to speak then there would have to be a great many classes. Large classes do not however rule out this approach, they merely present some logistical challenges. Mossa also suggests that it worked well in the example presented in part because the class met infrequently (once a week) and in part because formal assessment of the presentation was not required as peer pressure was a sufficient incentive. Newnham stresses the importance of encouraging student participation through a presentation at the start of the course that explains the benefits accruing to students.

The presentation of two quite different approaches but with similar goals demonstrates the flexibility of lecture review. There are obstacles to be overcome but if students can be encouraged to participate then there would appear to be a number of advantages to be gained by all involved.

4.4   Case study: lectures and distance learning

Aims: To make lecturers available to students via television broadcasts, video cassettes and WWW pages and to investigate the usefulness of this approach.

Sources: This review is based on papers by Fox (1996) and Hellwege *et al.* (1996)

Context:

The growth in distance learning programmes across the world and the opportunities presented by the WWW means that the potential for students to view video, film and other types of imagery without direct teacher involvement has grown. This mode of teaching has been further encouraged through the promotion of self-tuition and student-centred learning. With this opportunity questions arise about the presentation and use of visual lecture materials especially when the teacher is remote (Rowntree, 1990). The use of film and video in lectures has been discussed in Section 3. Concern was expressed over the power of such media and the need to develop students' critical abilities when viewing such imagery. Rose (1996) has recently suggested how this can be advanced using small discussion groups. Students' use of the WWW has only just begun to receive similar critical evaluation.

Two examples are cited here. The first case study by Fox was taken from Canadian experience where there has been a tremendous growth in HE students numbers yet the size of the country presents a geographic barrier to participation. The particular example concerns a course in human geography consisting of 48 one-hour lectures, which was subsequently broadcast and distributed using video tapes for students off-campus. The second case study by Hellwege *et al.* comes from Australia and uses the WWW to reinforce materials utilised in geology lectures both to students on site and those at a distant campus where video is also distributed.

Implementation:

Using video and televised broadcasts in lectures

Lectures were presented in a 450-seat lecture theatre using standard audio-visual equipment. They were recorded onto video-tape for subsequent broadcast or distribution. The visual course content included maps, satellite imagery, photographs and tables of data. Non-standard equipment included an overhead television camera for the display of illustrations.

The lectures were followed-up on campus by taught discussion groups led by graduate students, using exercises provided. Those off-campus were mailed the exercises and could seek advice by telephone. An answering machine was dedicated to calls about the course. The course enrolment was 812 (295 off-campus).

Teaching of geology using the World Wide Web

All lecture materials were prepared using Microsoft PowerPoint. The full text of the lecture and all illustrations were made available via the WWW prior to the lectures.

Evaluation:

For both approaches there were a number of technical facilities that the lecturers had to become adept with. This has cost implications in terms of training and provision of facilities but here the evaluation is more concerned with students' learning experiences.

Fox reports the results of two types of evaluation: student performance, and a student course evaluation questionnaire. In student performance there was a 20% drop out rate (left the programme) for those watching televised lectures. This is reported by the author to be average for such classes but twice the loss found for the classroom-based group. The average grade earned was slightly higher (64.9%) for those watching at home compared with an average of 62.7% for those on campus. It is difficult to draw firm conclusions from such data because of the different compositions and study regimes of the two groups.

Students noted the following advantages for using video:

- Can watch at their own pace, pausing and rewinding thus obtaining better lecture notes.

- Can watch when it suits and when the student has best concentration.

- Can watch in an environment free from distractions.

- The sound and visual presentation can be clearer/better on the video than in a large lecture theatre.

An advantage not noted in this study is that for foreign students the ability to play-back can help their understanding of both the subject and their English. It could also be argued that more eminent academics could be persuaded to teach on a course if the timing of its recording is determined by their own diaries.

The disadvantages included:

- The lack of contact with course teachers, especially the inability to ask immediate questions and receive immediate responses.

- Classroom questions and discussions may not be audible to those watching on television.

- Lack of direct classroom involvement may lead to inattention.

- Availability of the video cassettes leads to procrastination.

- The timing of the broadcast was important for those without a VCR.

- The presence of cameras in the classroom could be intimidating for students and staff.

The use of WWW pages appeared to make students more attentive in lectures as it reduced the need for note taking. Other advantages are a clearer structure and the same lecture can be repeated by a number of staff as the materials are already organised. It is reported that overall students felt their understanding and retention had improved.

'This is certainly the way to go. It gives the student time to ingest the information at a pace conducive to retaining the information.'

(student evaluation response, Hellwege et al. p.4)

Summary:

These case studies have highlighted examples of distance learning which is normally undertaken for particular geographical, economic or social reasons. There are however a number of lessons to be drawn from the experience. There are many advantages to be gained for students with learning difficulties if the lecture can be videoed and made available, maybe by placing it on the WWW. Students who are not being taught in their native language and those with disabilities can benefit greatly. There are significant costs in the production of such material in time and facilities. The lecturer also needs to decide whether a student's time is best spent watching and/or listening to a copy of the lecture or following up through reading or some other activity.

4.5   Case study: enhancing lectures by student interviews

A major obstacle to student participation is their lack of confidence. This is noticeable in first year courses and those dealing with topics not covered in pre-university education such as the development of geographical thought. Over fifteen years ago Cosgrove (1981) suggested the use of student interviews to tackle this problem, for a final year geographical concepts course.

Small groups of students were assigned a member of staff who then arranged an informal interview. Some recent publications and a CV were provided beforehand. The interview was used by the students to chart and understand the geographical career of the member of staff thus illuminating the philosophical and methodological debates within geography. It was concluded by Cosgrove that the approach generated student interest and enthusiasm but concern was expressed over how long staff would be prepared to accept being interviewed each year. In answer to this concern, this approach has been employed at UCL for a first year geographical ideas course for over the last 15 years. The following material is based upon the UCL experience:

The Objectives for Students

- To learn about the research aims, methods and ideas of a member of academic staff.

- To discover more about the relationships between geographical research and teaching.

- To develop group and interview skills.

- To exercise critical powers through reading the materials supplied.

The Method

1.   Groups of 4 to 5 students are allocated a member of academic staff.

2.   Each member of staff provides copies of three pieces of writing which are representative of their work. One of these is an unpublished manuscript. A CV may also be provided.

3.   A date is arranged for an interview.

4.   Before the interview the papers are read critically, using tutorial meetings for discussion of the material and planning the interview. Advice is given by the tutor on how to organise the interview, whether or not to tape it, and how to use open questions.

5.   The interview takes place, normally for about one hour.

6.   A report is prepared by each student individually which should address three questions:

a)   From your critical appraisal of the three pieces of written work, what are the objectives of the interviewee's research?

b)   How does the interviewee's present research relate to his/her earlier studies?

c)   How does the interviewee's research relate to his/her teaching, outside interests and concept of geography?

Evaluation:

The approach has stood the test of time in that staff have been willing to support the exercise for more than 15 years. The interviews are often great fun and challenging for staff and students, especially when trying to recall why a certain piece of research was undertaken.

The interview exercise has not been evaluated for some time and it is planned with a new first year curriculum that this will be addressed in 1997. The lecture course 'Ideas in Geography', in which the interviews take place, is assessed annually. In 1996 the overall student assessment was favourable, but with some reservations:

Very Poor	Poor	OK	Good	Very Good
5%	8%	30%	42%	16%

Most of the critical evaluation concerned the lecture content with students noting that they had not done anything about geographical ideas before coming to University and found the material difficult. There were no comments about the interviews, good or bad.

Contact:

For further information contact Dr. Clare Dwyer, Department of Geography, UCL (cdwyer@geography.ucl.ac.uk)

4.6   The use of case studies: the teaching and the learning of economic theory

Aims: Students need to be provided with increased opportunities to apply their knowledge, in the context of this example to 'think like economists', and the case study is one technique for achieving this aim.

Source: Carlson & Schodt (1995)

Context:

Case studies provide information and realistic examples of the need for problem solving but not the analysis. Students then have to frame pertinent questions, decide which analytical tools and principles apply and search for additional information. The emphasis is then upon 'doing' and developing creative skills. Although the case may be prepared individually or collectively the 'problem' is solved through class discussion with only guidance provided by the lecturer. This approach is then based on class group work and this has been employed with class sizes in excess of one hundred students.

Implementation:

Students are required to prepare a case study and to bring this to the class. This may involve a written analysis. In the in-class discussion can employ a range of strategies including role playing, courts of inquiry and so on (see Gibbs & Jenkins, 1992). It is important that the students be allowed the freedom to determine their own solution although the lecturer retains responsibility for involving all students and for ensuring important factual or analytical issues are addressed. Two examples are provided here:

A course on 'Development Economics', with junior and senior majors in economics, 20 to 25 students attending 60 minute classes held three times a week for 38 sessions. Six cases were used over ten sessions; topics included on economic growth of Brazil and development strategy in Singapore. Prior to each case all students wrote a four page response to a series of questions which, along with class discussions, accounted for 30% of the course assessment.

A course on 'International Monetary Problems', with senior economics majors and graduate students, 15 to 25 students, attending 75 minute classes two times a week for 30 sessions. Eight class sessions were devoted to cases. Most cases, such as how the EU should proceed to monetary union, were preceded by a written exercise handed in on the day the case was discussed.

In both examples lectures remained an important aspect of teaching but no longer dominated the instruction.

Evaluation:

Student evaluations showed that they believed the use of cases had contributed to their learning and none felt they should be replaced by lectures. Students noted in particular that cases helped to:

- illustrate the practical application of theories;
- make the class more interesting;
- clarify the subject structure;
- provide a better sense of the work of an economist.

Case studies also avoid the artificiality often found with the use of problem sets. There are however some issues that need to be acknowledged. In the teaching of economic theory case studies illustrate and challenge but they do not provide the theory or models that need to be understood. Lectures remain then an important aspect of this strategy. There is also a need to give up topics to make way for the case studies while it takes time to prepare each new case study although this may not be more than for a new lecture. Carlson & Schodt believe that the use of case studies is worth the effort, it is highly rewarding for both student and teacher and does much more than stimulate lively discussion. This approach is supported for teaching geography students by Grant (1997) who lists sources for a number of available cases (Table 3).

Table 3: *Sources of available WWW case studies (from Grant, 1997)*

Organisation	URL	Topics
CaseNet	http://csf.colorado.edu/CaseNet/index.html	Development, environment, foreign policy, history, international political economy, international trade and economics
The CHANCE Project	http://www.dartmouth.edu/~chance	Statistics, probability
Clearing House for Decision Case Education	http://www.decisioncase.edu	Sustainable agriculture, education technologies, extension education
The European Case Clearing House	http://www.ecch.cranfield.ac.uk/	Accounting, control and business environment, finance, policy and general management, human resources and organisational behaviour, marketing, production and operations management
Harvard Business School	http://www.hbsp.harvard.edu	Business, government and the international economy, competition and strategy, information, organisation and control systems, management policy, managerial accounting, marketing, organisational behaviour, technology and operations management

Conclusion

This Guide suggests that the traditional lecture, a monologue for 50 minutes, achieves at best very modest student learning and at worst leads to student disinterest with an emphasis upon content and factual regurgitation. We have indicated means for improving lectures even in their traditional form, but feel that real improvement will lie in greater student participation in lectures. A number of different approaches have been suggested and some have been reviewed. There remains however a number of obstacles to be overcome. Some fifteen years ago Gibbs (1982) listed a number of reasons why lectures remain the most prevalent teaching approach in HE and you may like to consider whether these still apply to you or to your institution today.

Here are some reasons why it is difficult to change teaching methods. Which reasons apply to you?

- Ignorant of the alternatives
- Do not know how effective lectures are
- Overworked
- Changes take too much of your time
- The alternatives to lectures appear to involve too much additional work
- There is a lack of resources
- The institution supports lectures, e.g. through allocation of teaching loads
- External and internal course validation supports lectures

There are a number of alternative approaches to teaching, which do not use lectures, and you may like to consult other Guides in this series and consider strategies such as student self-learning, student-led tutorials, computer-assisted learning and audio-tutorial laboratories to list but a few. Having decided to improve your lectures there is no lack of advice and we have tried to summarise geographical experience in this field; at the end of this Guide further sources of information are listed. In the end we agree with Anderson (1994) that the key to achieving greater student learning is for the lecturer to talk less.

6 References

6.1 References cited in the text

Anderson, L.W. (1994) *Lecturing to Large Groups* (Birmingham: Staff and Education Development Association Paper 81).

Birnie, J. & Mason O'Connor, K. (1998) *Practicals and Laboratory Work in Geography* (Cheltenham: Geography Discipline Network, CGCHE).

Bishop, M.P., Shroder, J.F. & Moore, T.K. (1995) Integration of computer technology and interactive learning in geographic education, *Journal of Geography in Higher Education,* 19(1), pp.97-110.

Bligh, D.A. (1971) *What's the use of lectures?* (Exeter: D. Bligh).

Bliss, J. & Ogborn, J. (1977) *Students' Reactions to Undergraduate Science* (London: Nuffield Foundation by Heinemann).

Brown, G. & Atkins, M. (1988) *Effective Teaching in Higher Education* (London: Methuen).

Brown, G. & Bakhtar, M. (1988) Styles of lecturing: a study of implications, *Research Papers in Education*, 3(2), pp.131-153.

Brown, S. (1997) The art of teaching small groups, *New Academic*, 6(1), pp.3-6.

Carlson, J.A. & Schodt, D.W. (1995) Beyond the lecture: case teaching and the learning of economic theory, *Journal of Economic Education* (winter), pp.17-28.

Chalkley, B. (1996) Geography and teaching quality assessment: how well did we do?, *Journal of Geography in Higher Education*, 20(2), pp.149-58.

Charman, D.J. & Fullerton, H. (1995) Interactive lectures: a case study in a geographical concepts course, *Journal of Geography in Higher Education*, 19(1), pp.57-68.

Clark, G. & Wareham, T. (1998) *Small-group Teaching in Geography* (Cheltenham: Geography Discipline Network, CGCHE).

Cryer, P. & Elton, L. (1992) *Active Learning in Large Classes and with Increasing Student Numbers* (Sheffield: CVCP Staff Development Unit).

Cosgrove, D. (1981) Teaching geographical thought through student interviews, *Journal of Geography in Higher Education*, 5(1), pp.19-22.

Cox, B. (1994) *Practical Pointers for University Lecturers* (London: Kogan Page).

Crewe, L. (1996) Graduate teaching assistants' training programmes: the challenge ahead, *Journal of Geography in Higher Education*, 20(1), pp.83-88.

D'Allura, J.A. (1991) Interactive education in an introductory environmental geology course, *Journal of Geological Education*, 39, pp.279-283.

Elton, L.R.B. & Laurillard, D.M. (1979) Trends in research on student learning, *Studies in Higher Education*, 4(1), pp.87-96.

Entwistle, N. (1984) Contrasting perspectives on learning, in: F. Marton, D. Hounsell, N. Entwistle (Eds.) *The Experience of Learning*, pp.1-18 (Edinburgh: Scottish Academic Press).

Flowerdew, R. & Lovett, A. (1992) Some applications of Murphy's Law using computers for geography practical teaching, *Journal of Geography in Higher Education*, 16(1), pp.37-44.

Fox, M.F. (1996) Teaching a large enrolment, introductory geography course by television, *Journal of Geography in Higher Education*, 20(3), pp.355-366.

Gardiner, V. & D'Andrea, V. (1998) *Teaching and Learning Issues and Managing Educational Change in Geography* (Cheltenham: Geography Discipline Network, CGCHE).

Gibbs, G. (1981) *Teaching Students to Learn* (Milton Keynes: The Open University Press).

Gibbs, G. & Jenkins, A. (1984) Break up your lectures: or Christaller sliced up, *Journal of Geography in Higher Education*, 8(1), pp.27-39.

Gibbs, G. & Jenkins, A. (Eds.) (1992) *Teaching Large Classes in Higher Education* (London: Kogan Page).

Gibbs, G., Haigh, M. & Lucas, L. (1996) Class size, coursework, assessment and student performance in geography: 1984-94, *Journal of Geography in Higher Education*, 20(2), pp.181-192.

Gold, J.R., Jenkins, A., Lee, R., Monk, J., Riley, J., Shepherd, I.D.H. & Unwin, D.J. (1991) *Teaching Geography in Higher Education: a manual of good practice* (Oxford: Basil Blackwell).

Gold, J.R., Revill, G. & Haigh, M.J. (1996) Interpreting the dust bowl: teaching environmental philosophy through film, *Journal of Geography in Higher Education*, 20(2), pp.209-22.

Goodland, S. (1997) Teaching assistants, *Studies in Higher Education*, 22(1), pp.83-92.

Gould, P. (1994) Perspectives and sensitivities: teaching as the creation of conditions of possibility for geographic thinking, *Journal of Geography in Higher Education*, 18(1), pp.277-89.

Grant, R. (1997) A claim for the case method in the teaching of geography, *Journal of Geography in Higher Education*, 21(2), pp.171-185.

Gunter, M.E. (1991) In class computer demonstrations for physical geology, *Journal of Geological Education*, 39, pp.373-375.

Hartley, J. & Davies, I. (1978) Note taking: a critical review, *Programmed Learning and Educational Technology*, 15(3), pp.207-224.

Hay, I. (1994) Notes of guidance for prospective speakers, *Journal of Geography in Higher Education*, 18(1), pp.57-65.

Healey, M. (1997) Geography and education: perspectives on quality in UK higher education, *Progress in Human Geography*, 21(1), pp.97-108.

Hellwege, J., Gleadow, A. & McNaught, C. (1996) Paperless lectures: an evaluation of the educational outcomes of teaching geology using the Web, *GEOCAL*, 15(Dec.), pp.3-6.

Higher Education Funding Council for England (1995a) *Report on quality assessment 1992-1995* (Bristol: HEFCE).

Higher Education Funding Council for England (1995b) *Subject overview report: quality assessment of geography 1994-1995* (Bristol: HEFCE).

Hodgson, V. (1984) Learning from lectures, in: F. Marton, D. Hounsell, N. Entwistle (Eds.) *The Experience of Learning*, pp.90-102 (Edinburgh: Scottish Academic Press).

Horton Smith, D. (1996) Developing a more interactive classroom: a continuing odyssey, *Teaching Sociology*, 24, pp.64-75.

Isaccs, G. (1994) Lecture practices and note taking purposes, *Studies in Higher Education*, 19(2), pp.203-216.

Jenkins, A. (1992) Active learning in structured lectures, in: G. Gibbs, & A. Jenkins (Eds.) *Teaching Large Classes in Higher Education*, pp.63-77 (London: Kogan Page).

Jenkins, A. (1997) Twenty one volumes on: is teaching valued in geography in higher education?, *Journal of Geography in Higher Education*, 21(1), pp.5-14.

Jenkins, A. & Smith, P. (1990) Quality control in geography courses: the role and practice of Her Majesty's Inspectors in British higher education, *Journal of Geography in Higher Education*, 14(2), pp.123-35.

Jenkins, A. & Smith, P. (1993) Expansion, efficiency and teaching quality: the experience of British geography departments, *Transactions of The Institute of British Geographers*, 18(4), pp.500-15.

Jenkins, A. & Youngs, M. (1983) Geographical education and film: an experimental course, *Journal of Geography in Higher Education*, 7(1), pp.33-44.

Kies, J.K., Williges, R.C. & Rosson, M.B. (1997) Evaluating desktop video conferencing for distance learning, Computers Education, 28(2), pp.79-91.

Matthews, H. & Livingstone, I. (1996) Geography and lifelong learning. *Journal of Geography in Higher Education*, 20(1), pp.5-9.

Macdonald, R.H. & Conrad, S.H. (1992) Writing assignments augment learning in introductory geology courses, *Journal of Geological Education*, 40, pp.279-286.

McBroom, W.H. & Reed, F.W. (1994) An alternative to a traditional lecture course, *Teaching Sociology*, 22, pp.328-332.

McCarthy, A., Smuts, B. & Cosser, M. (1997) Assessing the effectiveness of supplemental instruction: a critique and a case study, *Studies in Higher Education*, 22(2), pp.221-231.

McCartney, K., Kimball, R. & Swetnam, J. (1992) Teaching physical geology and laboratory courses by interactive television, *Journal of Geological Education*, 40, pp.367-372.

Mossa, J. (1995) Topic synthesis: a vehicle for improving oral communication skills, comprehension and retention in higher education, *Journal of Geography in Higher Education*, 19(2), pp.151-58.

Newnham, R.M. (1997) Lecture reviews by students in groups, *Journal of Geography in Higher Education*, 21(1), pp.57-64.

Petch, J. & Reid, I. (1988) The teaching of geomorphology and the geography/geology debate, *Journal of Geography in Higher Education*, 12(2), pp.195-204.

Phillips, M. & Healey, M. (1996) Teaching the history and philosophy of geography, *Journal of Geography in Higher Education*, 20(2), pp.223-242.

Race, P. & Brown, S. (1994) *500 Tips for Tutors* (London: Kogan Page).

Rose, G. (1996) Teaching visualised geographies: towards a methodology for the interpretation of visual methods, *Journal of Geography in Higher Education*, 20(3), pp.281-294.

Rowntree, D. (1990) *Teaching Through Self Instruction* (London: Kogan Page).

Saroyan, A. & Snell, L.S. (1997) Variations in lecturing styles, *Higher Education*, 33, pp.85-104.

Shepherd, I. (1998) *Teaching and Learning Geography with Information and Communication Technologies* (Cheltenham: Geography Discipline Network, CGCHE).

Sumner, G. (1984) Video kills the lecturing star: new technologies and the teaching of meteorology, *Journal of Geography in Higher Education*, 8(2), pp.115-24.

Wright, N. & Cordeaux, C. (1996) Rethinking video conferencing: lessons learned from initial teacher education, *Innovations in Education and Training International*, 33(4), pp.194-202.

6.2   Lectures in geography: sources of information

There is a vast amount of information on geography lectures and geography teaching available through WWW sites. With respect to lectures these can be divided into those providing information on lecture content, and those with information on teaching in lectures or lecturing style.

Content:

A large number of lectures have been placed on the WWW ranging from full lectures, such as inaugural lectures, through to outline summaries, but in the case of the virtual university project based at Austin, Texas, a much greater array of teaching materials and ideas are presented in the Virtual Geography Department, http://www.utexas.edu/depts/grg/virtdept/contents.html. Here are some other examples:

Geography lecture outline

"A stumble may prevent a fall": effects of technology on global change. Dr. Ruth DeFries, University of Maryland College Park. Copyright, 1996
http://www.geog.umd.edu/landcover/defries/geog123_5.html
size 2K - 29 Apr 96

Settlement geography of Eau Claire, Wisconsin

http://uts.cc.utexas.edu/~ikv/w367/367start.htm
size 2K - 6 Jun 96

Style:

There are also a number of WWW sites offering advice on lecture methods but few are from geography departments. For example:

Workshop: lecturing in a classroom

Teaching Assistant Fellow Program, a professional development program for experienced Northwestern University Graduate Teaching Assistants.
http://faber.ms.nwu.edu/faber_group/sanjeev_web/taf_web/lecturing.html

Advantages and disadvantages of the lecture method
> Office of Instructional Resources, University of Illinois at Urbana-Champaign
> http://www.oir.uiuc.edu/did/booklets/lecture/lecture1.html
> http://ulna.bio.psu.edu/GradStud/Krupnick/ws02a.htm

The pitfalls of lecturing: economic history services
> http://cs.muohio.edu/Archives/eh.teach/apr-95/0022.html

Case study abstracts of interesting and effective lecturing practice are available on the GDN WWW pages http://www.chelt.ac.uk/gdn.

Perhaps the best source of information for geographers interested in making changes to their teaching in lectures is the *Journal of Geography in Higher Education*. The following texts are also full of advice and suggestions:

Anderson, L.W. (1994) *Lecturing to large groups* (Birmingham: Staff and Education Development Association, Paper 81).

Cryer, P. & Elton, L. (1992) *Active learning in large classes and with increasing student numbers* (Sheffield: CVCP Staff Development Unit).

Cox, B. (1994) *Practical Pointers for University Lecturers* (London: Kogan Page).

Gibbs, G. & Jenkins, A. (Eds.) (1992) *Teaching Large Classes in Higher Education* (London: Kogan Page).

Gold, J.R., Jenkins, A., Lee, R., Monk, J., Riley, J., Shepherd, I.D.H. & Unwin, D.J. (1991) *Teaching Geography in Higher Education: a manual of good practice* (Oxford: Basil Blackwell).

Race, P. & Brown, S. (1994) *500 Tips for Tutors* (London: Kogan Page).